A NEW GENERATION OF ENGINE COMBUSTION PROCESSES FOR THE FUTURE?

From the same publisher

- **Direct Numerical Simulation for Turbulent Reacting Flows**
  T. Baritaud, T. Poinsot, M. Baum, Eds.
- **Multi-Dimensional Simulation of Engines Internal Flows**
  T. Baritaud, Ed.
- **Fuels and Engines. Technology. Energy. Environment (2 Volumes)**
  J.-C. Guibet
- **A new Generation of Two-Stroke Engines for the Future?**
  Proceedings of the international seminar held in Rueil-Malmaison, France, November 29-30, 1993
  P. Duret, Ed.
- **Lubricant Properties, Analysis, and Testing**
  J. Denis, J. Briant, J.-C. Hipeaux
- **Rheological Properties of Lubricants**
  J. Briant, J. Denis, G. Parc
- **Automobiles and Pollution**
  P. Degobert

IFP International Congress

A NEW GENERATION OF ENGINE COMBUSTION PROCESSES FOR THE FUTURE?

Proceedings of the International Congress
held in Rueil-Malmaison, France, November, 26-27, 2001

Edited by
PIERRE DURET
IFP

2002

Editions TECHNIP
27 rue Ginoux, 75737 PARIS Cedex 15, FRANCE

IFP International Congress

A NEW GENERATION OF ENGINE COMBUSTION PROCESSES FOR THE FUTURE?

Held in Rueil-Malmaison, France, November 26-27, 2001

Congress Secretariat

Ms Bettina CARUSO, IFP
92852 Rueil-Malmaison, France
E-mail: bettina.caruso@ifp.fr

Congress Chairman

Pierre DURET, IFP
92852 Rueil-Malmaison, France
E-mail: pierre.duret@ifp.fr

Technical Committee and Sessions Chairmen

Jean BOTTI, Delphi Automotive Systems
Benoit DELHAYE, PSA Peugeot-Citroën
André DOUAUD, IFP
Klaus FRÖHLICH, BMW
Kazumasa KATOH, Renault
Philippe PINCHON, IFP
Horst SCHULTE, FORD

Main authors

Robert M. SIEWERT, *GM R&D and Planning*
Costantinos VAFIDIS, *Fiat Research Centre*
Hiromichi YANAGIHARA, *Toyota Motor Europe*
Bertrand GATELLIER, *IFP*
Anders HULTQVIST, *Scania CV AB*
Bengt JOHANSSON, *Lund Institute of Technology*
Rudolf STANGLMAIER, *Southwest Research Institute*
Tomonori URUSHIHARA, *Nissan Motor Co., Ltd*
Jeff ALLEN, *Lotus Engineering*
Jacques LAVY, *IFP*
Tom MA, *Ford Motor Company*
Hua ZHAO, *Brunel University*
Andrew A. POURING, *Sonex Research, Inc*
Yoïchi ISHIBASHI, *Honda R&D Co., Ltd*
Peter WOLTERS, *FEV Motorentechnik GmbH*
Rainer AUST, *EUCAR SG Powertrain*
Jean DOREAU, *SAGEM*
Xavier MONTAGNE, *IFP*

Printed in France

ISBN 2-7108-0812-9

Authors addresses

ALLEN Jeff
LOTUS Engineering
Hethel, Norwich
NR14 8EZ
ENGLAND
JAllen@lotuscars.co.uk

AUST Rainer
DaimlerChrysler AG
HPC G252
D-70546 Stuttgart
GERMANY
rainer.aust@daimlerchrysler.com

BOTTI Jean
Delphi Automotive Systems
Centre technique - Paris
117, avenue des Nations
ZAC Paris Nord II - BP 60059
95972 Roissy CDG Cedex
FRANCE
jean.botti@dephiauto.com

DELHAYE Benoit
PSA Peugeot-Citroën
DRIA
Route de Gisy
78943 Velizy-Villacoublay Cedex
FRANCE
delhaye@calvacom.fr
delhaye@psinet.fr

DOREAU Jean
SAGEM
18, chaussée Jules César
BP 340 Osny
95526 Cergy-Pontoise Cedex
FRANCE
jean.doreau@sagem.com

FRÖHLICH Klaus
BMW AG
80788 München
GERMANY
klaus.froehlich@bmw.de

GATELLIER Bertrand
IFP
92852, Rueil-Malmaison, France
bertrand.gatellier@ifp.fr

ISHIBASHI Yoichi
Honda R&D Center
3-15-1 Senzul Asaka-shi Saitama
351-8555 JAPAN
yoichi.ishibashi@mail.a.rd.honda.co.jp

JOHANSSON Bengt
Division of Combustion Engines
LUND Institute of Technology
P.O. Box 118
S-22100 LUND
SWEDEN
bengt.johansson@vok.ith.se

KATOH Kazumasa
RENAULT
67, rue des Bons Raisins
92508 Rueil-Malmaison Cedex
FRANCE
kazumasa.katoh@renault.com

MONTAGNE Xavier
IFP
92852, Rueil-Malmaison, France
xavier.montagne@ifp.fr

POURING A.A
CEO
Sonex Research, Inc.
23 Hudson Street
Annapolis, Maryland
USA
sonex@erols.com

SCHULTE Horst
FORD Motor Co.
Suesterfeldstrasse 200
D-52072 AACHEN
GERMANY
hschult5@ford.com

SIEWERT Robert
GM R&D and Planning
Powertrain Systems Research Lab.
30500 Mound Rd.
Warren MI 48090-9055
USA
bob.siewert@gm.com

STANGLMAIER Rudolf
Engine Research Division
Southwest Research Institute
6220 Culebra Rd., Bldg. 151
San Antonio TX 78238
USA
rstanglmaier@swri.org

URUSHIHARA Tomonori
Powertrain Research Laboratory
Nissan Research Center
NISSAN MOTOR Co., Ltd.
JAPAN
tomonori-u@mail.nissan.co.jp

VAFIDIS Costantinos
Centro Ricerche Fiat
Strada Torino, 50
10043 Orbassano (TO)
ITALY
c.vafidis@crf.it

WOLTERS Peter
FEV Motorentechnik GmbH
Neuenhofstrasse 181
D-52078 AACHEN
GERMANY
wolters@FEV.de

YANAGIHARA Hiromichi
TOYOTA Motor Europe Marketing & Engineering
Advanced Technology Research Division
Hoge Wei 33 B
B-1930 Zaventem
BELGIUM
hiromichi.yanagihara@toyota-europe.com

ZHAO Hua
Brunel University
Department of Mechanical Engineering
Uxbridge, Middlesex,
UB8 3PH
UNITED KINGDOM
hua.zhao@brunel.ac.uk

Foreword

There is a growing interest and even some kind of effervescence in the new generation of engine combustion processes emerging from research and development work being done all over the world. These new combustion processes are generally named by various acronyms. Among them, HCCI for Homogeneous Charge Compression Ignition and CAI for Controlled Auto-Ignition seem probably the most appropriate to qualify respectively the diesel and gasoline applications.

It is well known that these HCCI and CAI (and others...) processes allow to drastically reduce the NOx and particulates emissions down to ultra low or near zero levels, while maintaining or even significantly improving the fuel economy.

Thanks to its more than 15 years' involvement and background in gasoline CAI and to its more recent involvement in diesel HCCI, IFP has in the last few years moved on from a pioneering role to worldwide recognition as a leading expert in this field of new engine combustion processes.

It appeared therefore quite natural at IFP to think organizing the first international Congress entirely dedicated to this subject.

We hope it will give to all the delegates the opportunity to make a short break in this worldwide effervescence and to have the time to contribute and also take benefit from the three main objectives of this event:

- to provide a clear up-to-date overview of the best state of the art of the new combustion technologies developed worldwide,
- to give the opportunity of fruitful exchanges and technologies benchmark generating new ideas for further progress,
- to enlarge the subject of the new engine combustion processes to a more global vision of what could be the "engine and energy source" couple in the powertrains of the future.

Satisfying these three objectives, this Congress should then bring its contribution to the definition of future engine and powertrain technologies road maps in which the automotive industry, the heavy duty and small engine sectors, the OEM's suppliers, the fuel companies and the R&D organizations (like IFP) could be all together committed in the future.

The dates of November 26 and 27, 2001 were selected for this Congress, in order to take place just before the SETC'2001 (Small Engine Technology Conference, 28-30 November in Pisa, Italy). It thus offers the delegates the opportunity of a full week of conferences with, in addition, the possibility of nice sight seeing programs in Paris as well as in Tuscany. We do hope that you will appreciate this event that we are very pleased to organize.

I personally wish to thank all the authors, speakers and session chairmen who accepted with enthusiasm our invitation to contribute to the Congress. The quality of their contribution combined with the active participation of the delegates will guarantee the success of this event.

To conclude, my great thanks go to my colleagues at IFP, especially to Mrs. Bettina Caruso and the Communication Division for the organization of the Congress, as well as to my colleagues of the "Engines and Energy" domain with special mention to Mr. André Douaud, Director of "Engines and Energy", to my secretary Mrs. Martine Gironce, and finally to my wife for her patience and support during the long hours of preparation work that were needed.

Pierre DURET
Congress Chairman

Table of contents

Congress contacts IV

Authors addresses V

Foreword VII

Keynote address
Pierre DURET 1

DIESEL HCCI COMBUSTION

Homogeneous-Charge Compression-Ignition (HCCI) of Gasoline-Like Fuels in a Single-Cylinder Diesel Engine
Robert M. SIEWERT 7

Variable Valve Actuation for In-Cylinder Charge and Combustion Control in D.I. Diesel Engines
Costantinos VAFIDIS, Antonio PICCININI, Laura GIANOLIO 25

Ignition Timing Control at Toyota "UNIBUS" Combustion System
Hiromichi YANAGIHARA 35

New Diesel Combustion Process to Achieve Near Zero NOx and Particulates Emissions
Bertrand GATELLIER, Bruno WALTER, Marjorie MICHE 43

GASOLINE CAI & HCCI COMBUSTION

The Heterogeneous Combustion Process in an HCCI Engine. Optical Characterization of Fluctuations and Length Scales
Anders HULTQVIST, Magnus CHRISTENSEN, Bengt JOHANSSON 55

Fuel Introduction Strategies for Pre-Mixed Compression-Ignition Combustion
Rudolf H. STANGLMAIER, Thomas W. RYAN III and Darius MEHTA 69

Parametric Study of Gasoline HCCI with Various Compression Ratios, Intake Pressures and Temperatures
Tomonori URUSHIHARA, Koji HIRAYA, Akihiko KAKUHOU, Teruyuki ITOH 77

Advanced Combustion Using a Lotus Active Valve Train. Internal Exhaust Gas Recirculation Promoted Auto-Ignition
Jeff ALLEN, Don LAW 85

Controlled Auto-Ignition (CAI): A New Highly Efficient and Near-Zero Nox Emissions Combustion Process for Gasoline Engine Application
Jacques LAVY, Jean-Charles DABADIE, Pierre DURET, Christian ANGELBERGER, Jean-François LE COZ, Jérôme CHEREL 101

Experimental Investigation of Controlled Auto-Ignition (CAI) Combustion in a 4-Stroke Multi-Cylinder Gasoline Engine and Drive Cycle Simulations
Tom MA, Hua ZHAO, Jian LI, Nicos LADOMMATOS 115

ACTIVATED RADICALS COMBUSTION PROCESSES

Radical Ignition in Low Compression Ratio D.I. Engines Enabled by Sonex Piston Design
David A. BLANK, Andrew A. POURING 127

Activated Radical Combustion in a High-Speed High-Power Pneumatic Direct Injection Two-Stroke Engine
Yoïchi ISHIBASHI, Kenji NISHIDA, Masahiro ASAÏ 141

Radical Activated Combustion. A New Approach for Gasoline Engines
Peter WOLTERS, Wolfgang SALBER, Jochen DILTHEY 153

WHAT ENGINE COMBUSTION PROCESSES FOR THE FUTURE?

The EUCAR Vision of the Future of Powertrains
Rainer AUST 165

Contribution of New Generation Variable Valve Trains and Engine Management Systems for the Control of New Combustion Processes
Jean DOREAU 171

What Will Be the Future Combustion and Fuel-Related Technology Challenges?
Xavier MONTAGNE, Pierre DURET 175

A New Generation of Engine Combustion Processes for the Future?
© P. Duret (Editor) and Editions Technip, Paris, 2001, pp. 1 - 4
27 rue Ginoux, 75015 Paris

KEYNOTE ADDRESS

Pierre Duret
IFP, France

INTRODUCTION

The purpose of this International Congress is to provide an opportunity for experts from the automotive industry, the heavy duty and small engine sectors, OEM's suppliers, fuel companies and R&D organizations to exchange views on the chances of success of this new generation of engine combustion processes.

First of all, it is necessary to make an attempt of definition of this "new generation of engine combustion processes". What we call here "new" combustion processes are investigated for their potential to reduce to very low level the production of NOx and particulates. They constitute alternative or complementary answers to the sophisticated and complex after-treatments strategies. Homogeneous Charge Compression Ignition (HCCI) and a number of acronyms designate these combustion processes. Their principle consists in preparing an highly diluted mixture (not always homogeneous!), in achieving its simultaneous ignition in the whole space of the combustion chamber and in precisely controlling the combustion event for best performance in terms of efficiency and NOx/PM emissions (timing and heat release rate during the cycle). Because of the thermo-chemistry of the charge, the thermal NOx formation and the soot production are in principle much lower than in the diffusion flame typical of the conventional diesel or in the propagation flame typical of the conventional spark ignition engine.

Nevertheless, as will be shown in the first section of this paper, the interest for such combustion processes is not really new, since they have already been the subject of a great number of studies by worldwide researchers. However, as shown in Figure 1, a recent breakthrough can be observed in the research and development activity in this field during the last few years (first increase in 1996, with a sharp increase since 1999-2000), thanks to the new possibilities offered by recent engine technologies such as electronic control, direct injection, variable valve actuation, ...

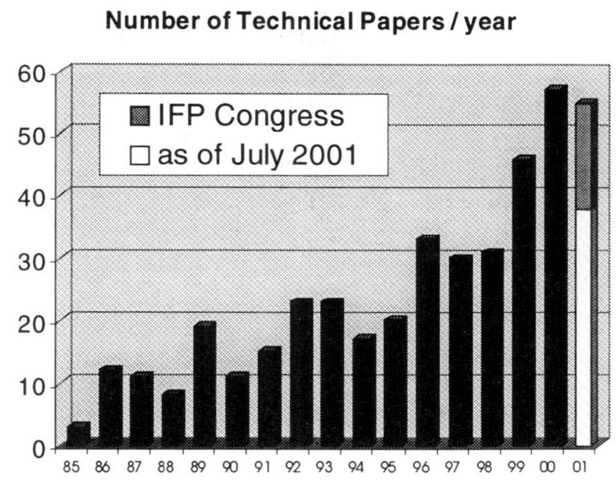

Figure 1: Statistical analysis of the number of technical papers related to "new" combustion processes and published from 1985 to 2001 (year 2001 is incomplete as indicated in the graph).

This is the reason why we thought it was now time for an open scientific and technical exchange about the chance of these "new generation" of combustion processes to be handled under the bonnet of a future passenger car.

A LONG ENGINE RESEARCH STORY

What we name "new" combustion processes have already been explored by several researchers, especially in Japan and partly in Europe and the USA. Various names and acronyms have been given in the past to these combustion processes, including ATAC for "Active Thermo Atmosphere Combustion" process (developed and industrialized during the seventies by Nippon Clean Engine for a small generator application), T.S. for "Toyota-Soken" combustion, A.R. for "Activated Radicals" combustion (developed and industrialized by Honda recently in motorcycle production in Japan as well as in the EU), or FDCCP for "Fluid Dynamically Controlled Combustion Process", CIHC for Compression-Ignited Homogeneous Charge Combustion, HCCI for Homogeneous Charge Compression Ignition or PCCI for Premixed-Charge Compression Ignition,...

To simplify, among all these acronyms, the generic term of HCCI is very often used. Nevertheless, the other generic term CAI for "Controlled Auto-Ignition" could be much more appropriate when the heterogeneity of the charge (of fuel, air and EGR) is one of the main characteristics of the combustion process.

Figure 2 shows some of the companies and universities which were the most active in this field during the last three years according to the criterion of the number of technical papers published.

THE "CO2-EFFICIENCY" VERSUS "NOX-PM" DILEMMA OF STRATIFIED COMBUSTION PROCESSES

In Europe, there is a strong concern about particulates and nitrogen oxides from heavy-duty (HD) as well as from light duty (LD) diesel engines because of their major contribution to environmental pollution.

Similarly, we can observe on the spark ignition (SI) engine side as used in LD vehicles, the arrival of a new generation of high efficient Direct Injected (DISI) gasoline engines. These new engines bring significant improvement in terms of CO2 emissions compared to that of stoichiometric SI gasoline with a three way catalyst. Nevertheless, this new generation of DISI engines also presents the same drawback of rather high NOx, and non negligible PM. Nevertheless, since this new generation of DISI engines is not yet produced in great number, its contribution to the NOx and PM pollutant problem is today rather small compared to that of the diesel engines. From the above comments, it clearly appears that the diesel as well as the DISI engines provide a good answer for low CO2, but presents the drawbacks of high NOx and PM that cannot be

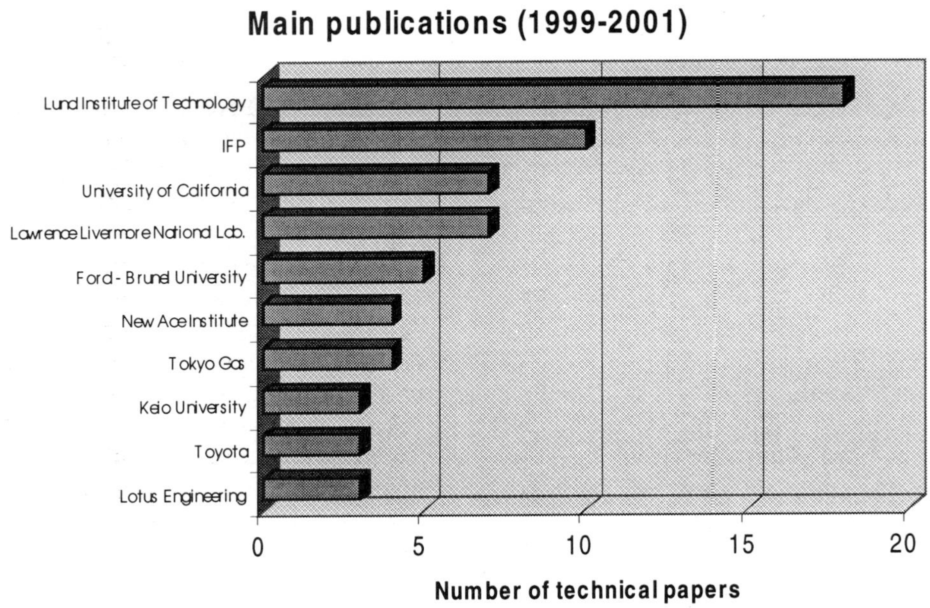

Figure 2: Main companies having published technical papers related to the new generation of engine combustion processes during the last 3 years (statistical analysis).

easily and efficiently treated with aftertreatment devices. This is the "CO2–efficiency" versus "NOx-PM" dilemma.

What characterizes both diesel and DISI engines and is mostly responsible of the high NOx-PM is the stratified mode of combustion. Engine experts very well know that stratified combustion generates soot and NOx emission. On the other hand, the current homogeneous combustion as used in gasoline passenger cars needs to be carried out at stoichiometry in order to obtain efficiency from the aftertreatment device.

IFP AT THE CENTER OF A LARGE EUROPEAN CLUSTER OF PROJECTS

One of the missions of IFP is to federate at an European level (involving car manufacturers, OEM's suppliers, R&D labs and universities) all the research and development of new generation engine combustion processes like gasoline CAI (Controlled Auto-Ignition) and diesel HCCI (Homogeneous Charge Compression Ignition), where the combustion happens simultaneously in the whole "SPACE" of the combustion chamber providing almost no NOx and particulates and the robustness of ultra low emission level during the whole vehicle life without needing sophisticated "fuel quality"-sensitive aftertreatment (at least for Euro 4).

This is the reason why IFP has set up and is now driving several large multi-partner European projects, all these projects being grouped within the "SPACE Combustion" cluster as shown in Figure 3.

The purpose of the "SPACE" cluster is to give an alternative solution either to the stratified conventional diesel or the DISI engines and to the gasoline homogeneous stoichiometric engine. The target of this alternative solution is to overcome the "CO2–efficiency" versus "NOx-PM" dilemma by the use of highly diluted Controlled Auto-Ignition (CAI) and Homogeneous Charge Compression Ignition (HCCI) in respectively gasoline and diesel engines. With this highly diluted homogeneous charge combustion, it is then possible to reach "CO2–efficiency" levels similar to the best stratified engines while maintaining ultra low "NOx-PM" without fuel sensitive aftertreatment. Since the purpose of the "SPACE" cluster is to develop in parallel and in synergy, the homogeneous highly diluted combustion solution for the gasoline (4-SPACE), the LD (SPACE LIGHT) and the HD (Hy-SPACE) diesel engines, as well as to develop the generic combustion process understanding and associated numerical simulation tools (PLANET), and the corresponding fuel specifications well adapted to take the greatest benefit of HCCI combustion (possible future SPACE FUEL project), this clearly demonstrate the problem solving approach adopted.

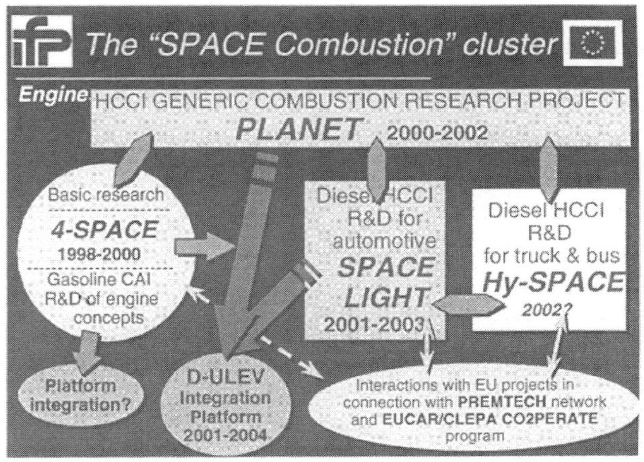

Figure 3: The European "SPACE Combustion" cluster driven by IFP

The following European projects currently belong to the SPACE Combustion cluster:

- The "4-SPACE" (4-Stroke Powered gasoline Auto-ignition Controlled combustion Engine) project was driven by IFP with DaimlerChrysler, Ford and PSA as main partners and with Brunel and Heidelberg Universities as associated partners.
Its main achievements were to demonstrate that CAI (Controlled Auto-Ignition) can be applied to automotive gasoline engines without any engine redesign except for Variable Valve Actuation, without increased compression ratio and without intake charge heating.
Thanks to the ultra low NOx and high fuel economy achieved, the 4-SPACE project is now followed by several industrial development projects for the application of the CAI combustion in future PFI (Port Fuel Injection) as well as in gasoline DI automotive engines that will be equipped either with mechanical, electro-hydraulic or electro-magnetic variable valve actuation systems.

- The "PLANET" (PLatform on Auto-ignition Numerical Engine simulation Tools) project is

driven by IFP with DaimlerChrysler, Fiat Research Center, PSA and Volvo as main partners and some universities as associated partners.
Its main purpose is to develop knowledge and CFD design tools to support the development of the HCCI engines. Its main deliverables are optical engines with associated diagnostics for HCCI studies, HCCI combustion data base, complex and simplified chemical kinetic schemes for HCCI, turbulent auto-ignition/combustion models for HCCI in order to be implemented in commercial CFD codes.

• The "SPACE LIGHT" (Whole SPACE combustion for diesel LIGHT duty vehicles) project is driven by IFP with OPEL, Renault, Lotus Cars and CRT (Common Rail Technologies) as main partners and Politecnico di Milano and Brunel University as associated partners.
Its main features are an upgraded Common Rail Direct Injection technology with appropriate "SPACE" spray pattern, a fully Variable Valve Actuation, a precise control of the charge mixing … in order to achieve a wide range of controlled Homogeneous Charge Compression Ignition.

• Finally, the possible new "Hy-SPACE" (Heavy duty diesel whole SPACE combustion) project will research and develop the application of HCCI combustion for diesel heavy duty truck and bus applications. It will include three major EU HD vehicles manufacturers, one major injector and EMS supplier, and one university.

All these projects are undertaken in close cooperation and networking with the EUCAR (EUropean Council of Automotive Research) organization and with interactions with other EU projects and technology platforms as shown in Figure 3, in order to guarantee the best industrial impact of these R&D results and achievements.

Being at the center of this European network of companies involved in these new combustion processes, it was therefore natural for IFP to think organizing this first international event entirely dedicated to this specific subject of growing interest.

CONCLUSION

To conclude this keynote address, I sincerely hope that this event will provide all the speakers and delegates with the following opportunities:

- To discuss how both gasoline CAI (Controlled Auto-Ignition) and diesel HCCI (Homogeneous Charge Compression Ignition) combustion processes could open a new scope for clean and efficient energy conversion by powertrains,
- To examine how the most advanced engine control technologies, such as VVA (Variable Valve Actuation), DI (Direct Injection),… will allow realistic practical applications of gasoline CAI / diesel HCCI in the near future production of powertrains,
- To convey the up-to-date development of the most advanced scientific tools such as optical access engines and associated diagnostics, and 3D CFD tools, allowing to earn the knowledge necessary to support the design and engineering process up to the industrial application,
- And finally, to open some discussions about the fuel related issues. If successful up to production, and thanks to their near zero raw emissions qualities, the CAI & HCCI combustion modes could bring about a complete revolution in the fuel technology. And why not, as shown in Figure 4 below, a future single fuel could be used in a unique "Combined Combustion System" derived from both gasoline CAI and diesel HCCI….

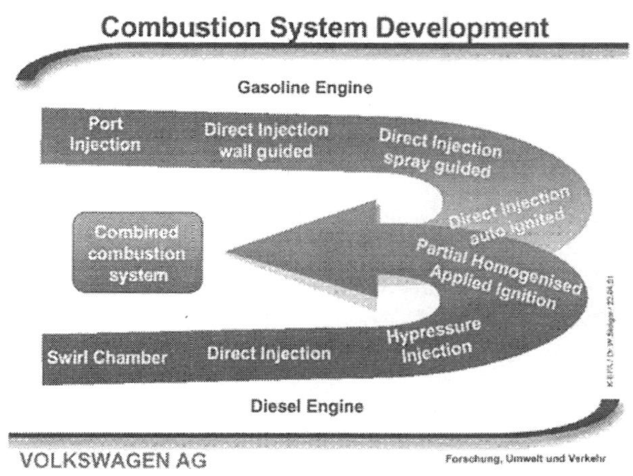

Figure 4: The idea of a unique fuel for a "Combined Combustion System" derived from both gasoline auto-ignition and homogeneous diesel combustion (source: Dr. Steiger, VW).

DIESEL HCCI COMBUSTION

A New Generation of Engine Combustion Processes for the Future?
© P. Duret (Editor) and Editions Technip, Paris, 2001, pp. 7 - 24
27 rue Ginoux, 75015 Paris

HOMOGENEOUS-CHARGE COMPRESSION-IGNITION (HCCI) OF GASOLINE-LIKE FUELS IN A SINGLE-CYLINDER DIESEL ENGINE

Robert M. Siewert
General Motors R&D and Planning, USA

ABSTRACT

A small-bore (76-mm) direct-injection single-cylinder diesel engine was used for evaluating homogeneous-charge compression-ignition (HCCI) of gasoline-like fuels. Fuel was not injected directly; rather, a pre-mixed charge of gasoline and air was prepared external to the engine to provide a well-vaporized mixture. Compression-ignition was achieved by independently varying equivalence ratio, EGR, and inlet temperature. Limited variations in speed, manifold pressure, and in-cylinder swirl also were made. Two gasoline's having a 91 research octane number (RON), n-heptane (0 RON), and iso-octane (100 RON) all were included. Most testing was at a compression ratio of 18.7:1, while a limited amount was done at a lower value of 16.1:1.

Selection of data with acceptably low NOx (under 1.0 g/kg fuel), smoke (at or near zero), and combustion noise (under 6000 kPa/ms maximum pressure rise) and all at near diesel thermal efficiency gives the range of load, EGR, and equivalence ratio for good HCCI combustion. With gasoline fuels, the load range was from about 100 to 600 kPa net mean effective pressure (NMEP). Equivalence ratio varied over a range from 0.2 to 1.0, EGR from 0 to 60%, and inlet temperature from 50 to 220°C.

INTRODUCTION

Homogeneous-charge compression-ignition (HCCI) combustion is currently a topic of intense, worldwide research due to its implications as a potential enabler to meet increasingly stringent exhaust emission requirements for internal combustion engines. Over the years it has been studied in many forms and considered for a variety of applications.

The author has first-hand experience with a variation of HCCI-combustion that dates back to 1970. The concept was called the "staged-combustion compound-engine", an engine configuration where the exhaust products from spark-ignition of a fuel rich mixture in one cylinder were passed on to a second-stage cylinder, after the addition of sufficient air for an overall lean mixture to complete combustion [1]. Compression-ignition of this homogenous charge occurred in the second-stage cylinder, producing about half of the power of the engine and resulting in very low NOx emissions due to the dilute nature of the mixture. Control of combustion in the second-stage cylinder was by thermal ignition and was accomplished by varying mixture temperature and/or compression ratio [1-3].

Thermal ignition, however, was not deemed suitable for following the transient operation required of an automotive engine. Rather, spark-ignition was then implemented for control of the second-stage combustion process. With sufficient inter-cooling to prevent thermal ignition and with sufficiently high fuel concentrations (primarily CO and H_2) for the second-stage mixture, spark-ignition was found to be an effective form of "forced ignition" that provided the needed control. Engine-out NOx emissions under 1.0 g/mi on the 1975 US FTP were demonstrated in an 2041-kg weight vehicle powered by an 7.46-L V-8 staged combustion engine which operated at an overall lean air-fuel ratio of 17:1 [4]. Although the fuel consumption was less than for the rich thermal reactor experimental vehicles that were being investigated at that time, the advent of the three-way catalyst enabled meeting the 1.0-g/mi NOx levels with good control of HC and CO and improved fuel economy. That new technology closed the chapter on any further work on the staged combustion engine.

In the late 1970's, a type of HCCI combustion called "Active Thermo-Atmospheric Combustion (ATAC)" within two-stroke engines was reported [5]. Due to the high residual temperatures and unburned fuel concentrations typical in two-stroke engines, under some circumstances auto-ignition of the charge would occur with very stable and more complete combustion than with spark ignition. More research was reported in the 1980's: studies on two-stroke engines continued and other work focused on the fundamentals of pre-mixed HCCI combustion [6-7] in four-stroke engines. Then from the 1990's and later, the citations in the literature have literally exploded, but their complete review is beyond the scope of this paper.

Some HCCI work has been reported for the four-stroke diesel engine using diesel fuel. The work generally falls into one of three methods of fuel introduction:

1. Pre-mixed fuel-air charge (including fumigation, port fuel injection or direct injection during the intake stroke)
2. Early direct injection of fuel (injection early in the compression stroke)
3. Late direct injection of fuel (injection near tdc but later than typical for conventional diesel engines)

Although pre-mixing of diesel fuel for HCCI combustion has been demonstrated [7-10], it requires very high mixture temperatures due to the low fuel volatility if one desires complete fuel vaporization.

Direct injection of diesel fuel during the compression stroke poses challenges for proper vaporization, mixing, and combustion [11-17]. That approach requires creative matching of the piston bowl shape and fuel injection systems. Recently, however, some promising results are given in [17].

Under some circumstances, late injection of diesel fuel when combined with large amounts of exhaust gas recirculation (EGR) can result in a form of HCCI that can lower smoke and particulate mater with relatively low NOx and a minimal fuel consumption penalty [18-20].

Another concern with using diesel fuel for HCCI combustion may be its tendency to ignite too early, although this might be an advantage for the early injection approach which appears to be the case for [17]. Some workers have lowered the Cetane number of the fuel for better ignition timing [10-13]. An alternative is to change the ignitability of the fuel by blending mixtures such as with primary reference gasoline fuels or gasoline and diesel fuel to control ignition timing [10].

For this study, we were interested in the possible application of HCCI combustion to direct-injection diesel engines. As a first step, an exploration of the operating range of such an engine with acceptable pre-mixed charge HCCI combustion was undertaken. Gasoline fuels were chosen due to their high volatility that simplified mixture preparation, thus ensuring a well-vaporized pre-mixed fuel-air mixture. The range of fuel ignitability included that of a high-octane gasoline (iso-octane), typical gasoline, and that approximating a high Cetane diesel fuel (n-heptane).

DESCRIPTION OF EXPERIMENTS

ENGINE AND DATA ACQUISITION - The engine used in these experiments was a single-cylinder direct-injection diesel engine as described in [18] and having a bore of 76-mm and a stroke of 94-mm. This DOHC four-valve engine had port throttles in each intake port to alter in-cylinder swirl. The design compression ratio was 18.7:1. However, some additional tests were conducted with a reduced compression ratio of 16.1:1.

For the HCCI experiments reported in this paper, direct-injection of diesel fuel was not used. Rather, the engine was supplied with a pre-mixed charge of gasoline, air and EGR. Fuel was injected onto a hot plate located in a large inlet tank that received both heated air and recirculated exhaust gas. The intent was to provide a well-mixed and fully vaporized mixture of fuel, air, and recirculated exhaust to the inlet ports.

HCCI combustion was achieved by controlling the inlet mixture temperature, inlet mixture pressure, fuel quantity, and EGR quantity.

Engine speed was controlled by an engine dynamometer and data were obtained at steady-state operating conditions. Low-speed data acquisition included the customary measurements common with engine testing including fuel, air, and EGR rates, temperatures, pressures, and engine-out exhaust gas analysis. Exhaust smoke values also were recorded. Cylinder pressure data were obtained at one-half degree increments for 50 consecutive cycles. Post-processing of that data included a heat release analysis.

TEST FUELS - The four test fuels used in this study are given in Table 1.

Table 1 - Test Fuels Used in Study.

Fuel	RON
Iso-Octane	100
Gasoline A	91.8
Gasoline B	91.4
n-Heptane	0

Gasoline fuels were chosen because of their relatively low boiling point and ease of vaporization. The range of research octane number (RON) was from 0 to 100.

HCCI DATABASE AND SELECTION PROCESS

A database (in Microsoft Excel) was accumulated for HCCI tests of each fuel listed in Table 1 over a wide variety of operating conditions. Many of the tests were quite exploratory in nature and resulted in HCCI combustion that would not be considered acceptable in terms of exhaust emissions, thermal efficiency, or noise. For example, single-variable test series such as changing inlet temperature might include over-advanced combustion with high NOx emissions and excessive combustion noise and low thermal efficiency for high values of inlet temperature. For low values of inlet temperature, combustion retardation, stability, and efficiency all might be unacceptable. Within such a test series, there might be just one or a few tests considered acceptable in terms of combustion phasing, emissions, efficiency, and noise.

It was of interest to select from that database points having acceptable HCCI combustion according to the criteria outlined in Table 2. The variables chosen were NOx emissions, indicated thermal efficiency (based on dynamometer firing and motoring torque measurements), maximum cylinder pressure rise, and coefficient of variation (COV) of indicated mean effective pressure (IMEP). Near zero smoke also was desired. However, due to the very low smoke levels for all the tests, it was not necessary to include that variable.

Table 2 - Criteria for selection ("filtering") of data.

Variable	Units	Criteria	Value
NOx EI	g/kg	<=	1.0
ITE (dyno indicated)	%	>=	40
Max. Pres. Rise	kPa/ms	<=	6000
COV IMEP	%	<=	5

Selection of the variables and establishing the selection value was based on the judgment of the author. The intent was to select data with "acceptable" emissions, efficiency, and noise for HCCI combustion. The accompanying operating conditions for that data would then define the range for acceptable HCCI combustion.

The selection process used is referred to as "filtering" only because the "filter" function of Microsoft Excel was employed to enter the selection criteria. In presenting the results, the "unfiltered" data are compared side-by-side with the "filtered" data for each variable of interest over the range of net mean effective pressure (NMEP). NMEP, the sum of indicated mean effective pressure (IMEP) and pumping mean effective pressure (PMEP), is determined from cylinder pressure measurements and is computed over the entire engine cycle.

RESULTS

EXHAUST EMISSIONS - NOx emission index (NOxEI) versus NMEP is given in Figures 1 with the filtered data shown on the left side and all the data on the right side.

As specified in Table 2, the filtered NOx levels were equal or less than 1.0 g/kg. For comparison, the same maximum scale is used for all the data. Note that for all the data, some extremely high levels of NOx EI were reached (i.e. >100 g/kg), emphasizing that low NOx levels are by no means automatic with HCCI combustion.

HC and CO emissions are given in Figures 2 and 3, respectively. The HC and CO ranges and magnitudes are quite high, particularly at the lower NMEP's. Filtering reduced the magnitude and range of HC and CO somewhat. Ranking HC and CO by fuel type, n-heptane typically had the lowest values and iso-octane had the highest values at a given NMEP.

Since the diesel engine combustion chamber was not designed for a homogeneous charge, it is not ideal for minimizing the crevice volume that can be one source of HC emissions. For this particular engine, that volume was about 4.5% of the clearance volume. That would be equivalent to an HC emission index of 45 g/kg if none of the fuel in that crevice burned. Put another way, this would be equivalent to throwing away 4.5% of the fuel, an obviously undesirable situation for both emissions and efficiency.

Smoke emissions are given in Figure 4. Exhaust smoke was very low, with the filter smoke number (FSN) typically less than 0.1 for all but a few points.

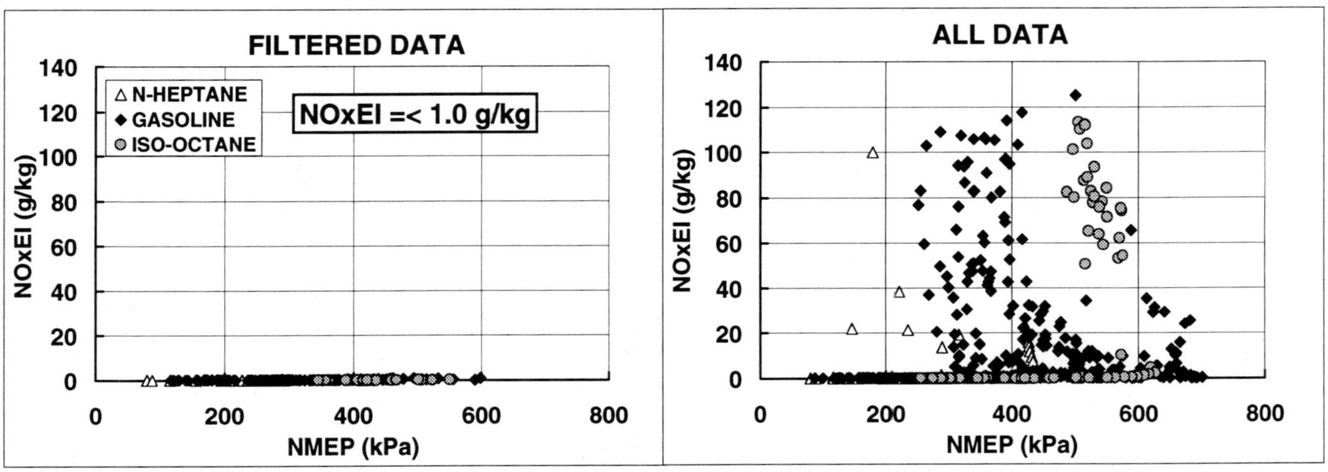

Figure 1 – NOx Emission Index vs NMEP for filtered data (left) and all data (right).

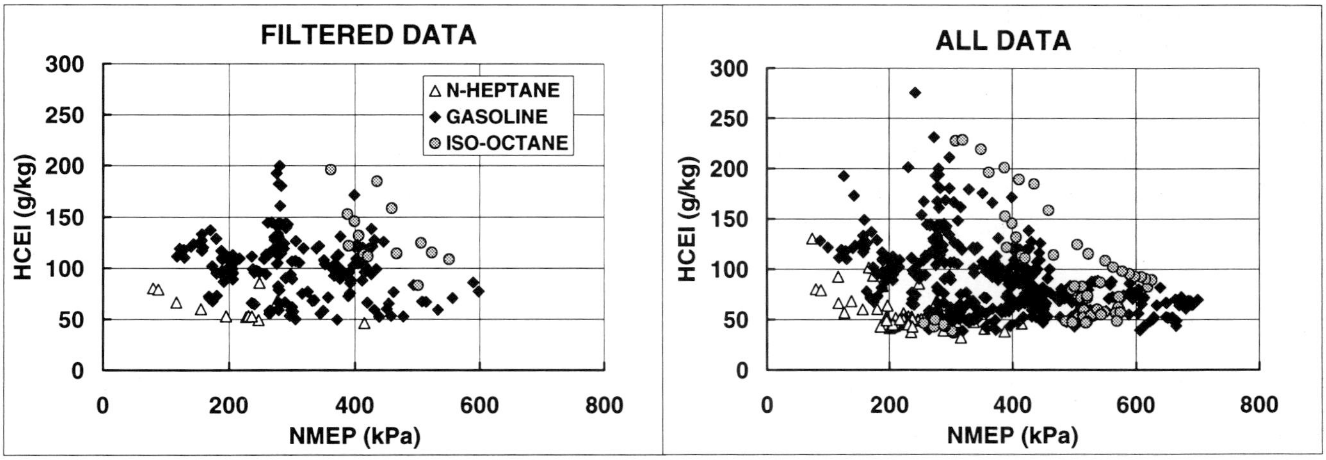

Figure 2 – HC Emission Index vs NMEP – for filtered data and unfiltered data.

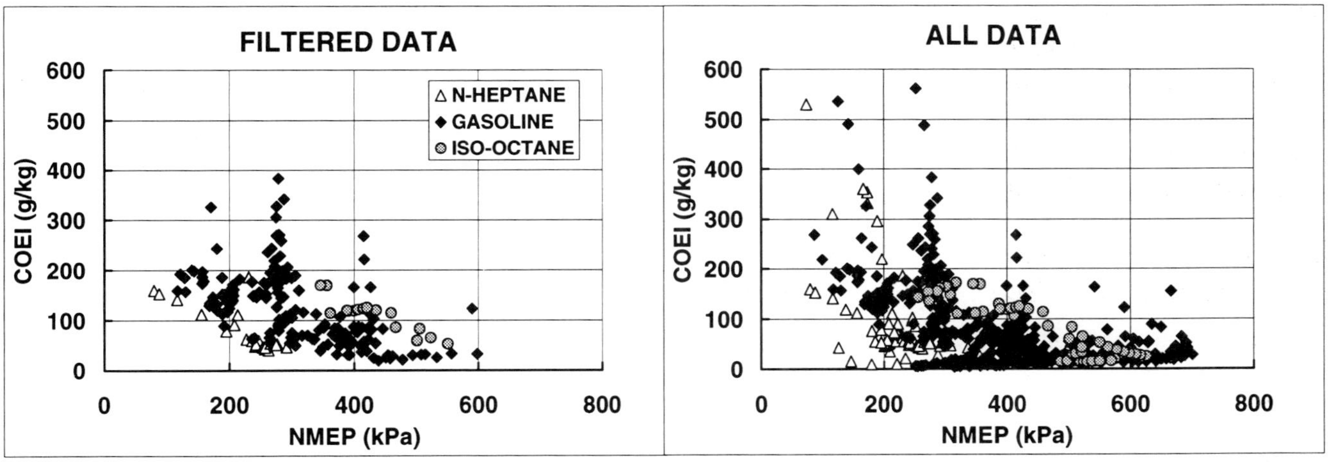

Figure 3 – CO Emission Index vs NMEP – for filtered data and unfiltered data.

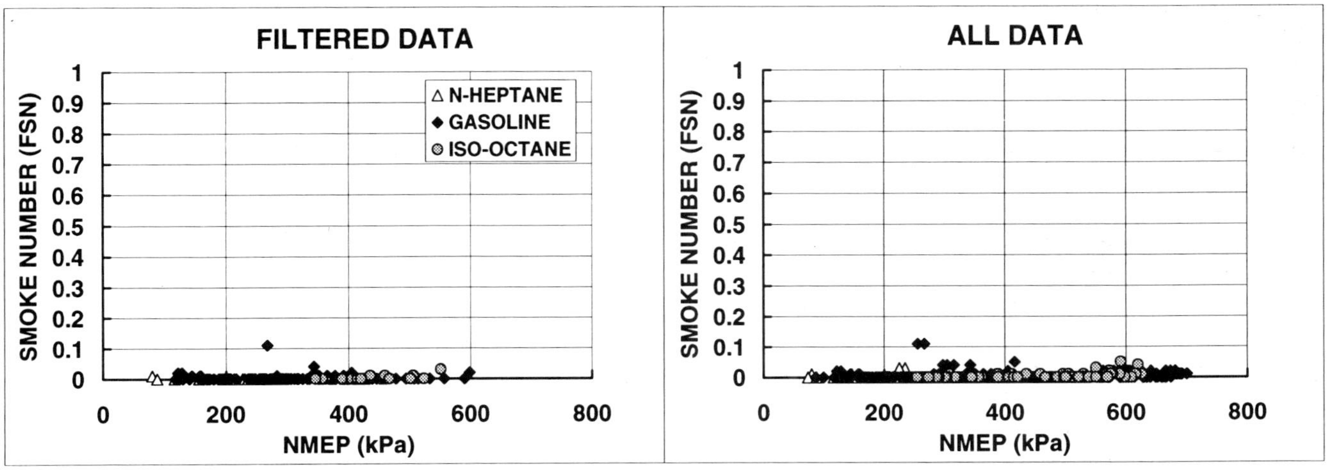

Figure 4 – Smoke number vs NMEP – for filtered and unfiltered data.

THERMAL EFFICIENCY AND COMBUSTION EFFICIENCY

Indicated thermal efficiency, based on dynamometer firing and motoring torque measurements (ITEdyno), is given in Figure 5. The torque-based measure of efficiency was chosen for filtering (Table 2) rather than cylinder-pressure-based efficiency due to the more uniform values over the load range. Thus a single filtering value (40% in Table 2) was quite effective in selecting data over the entire load range.

Maximum values of ITEdyno are in the mid-40's over the range of NMEP's covered. Filtering had little effect on the maximum values, but eliminated all values less than 40%.

The range of net indicated thermal efficiency (NITE) based on cylinder pressure measurements is given in Figure 6. The magnitudes of NITE are less than ITEdyno due to the differences in the underlying measurements of mean effective pressure. NMEP is always less than IMEPdyno, particularly at lower MEP's. Thus, NITE characteristically falls off at lower NMEP's as shown in Figure 6. For the filtered data, NITE varies from about 42% at the highest NMEP to less than 20% at the lowest NMEP.

The maximum thermal efficiencies shown in Figures 5 and 6 approach those of a diesel engine.

Combustion efficiency is given in Figure 7. Maximum values of combustion efficiency exceed 95%, Figure 7, and decrease slightly for the filtered data. Minimum values increased from about 60% to 75% for the filtered data. Higher magnitudes are associated with more advanced combustion that occurs with the more ignitable fuels (having lower RON's.) HC and CO emissions and computed H_2 determine the combustion efficiency. Lower combustion efficiency limits the maximum thermal efficiency possible.

A good direct-injection diesel engine will have combustion efficiencies in the range of 98 to over 99%.

COMBUSTION

START OF COMBUSTION AND COMBUSTION PHASING – Start of combustion, as indicated by the location of the 10% burn point from the heat release analysis, is given in Figure 8.

Start of combustion varies from about –20 to about +15 datc. Normal heptane clearly ignites the earliest of the fuels tested and iso-octane generally ignites the latest. Filtering reduced the range of SOC for each fuel, with gasoline falling mostly between 0 and 10 datc.

The phasing of combustion is given by the location of the 50% burn point, Figure 9. For gasoline, combustion phasing predominately lies between 10 and 20 datc for the filtered data. The trends with fuel follow those seen for SOC, Figure 8. Combustion phasing has a significant impact on both thermal efficiency and emissions. Note how over-advanced combustion is with n-heptane

COMBUSTION DURATION - Combustion duration's are given in Figures 10 and 11 for 10 to 75% burn and for 10 to 90% burn, respectively. The 10 to 75% burn duration's are relatively short and most are less than 60°. Filtering has eliminated duration's less than about 20°. However, the nature of the heat release is for an initial fast burning followed by a slower burning toward the end of heat release. That lengthens the 10 to 90% combustion duration considerably (Figure 11) as compared to the 10 to 75% duration (Figure 10). For the filtered data, the range of 10 to 90% burn is from about 60 to 120°, and the latter is long enough to have an negative effect on thermal efficiency.

BURN RATE, CYLINDER PRESSURE DEVELOPMENT, AND NOISE – Maximum burn rate, given in Figure 12, impacts cylinder pressure development and combustion noise. Maximum burn rate varies over a wide range for all the data given in Figure 12, but is greatly reduced for the filtered case. There, peak values have been reduced from over 50 to less than about 15 %/° crankangle.

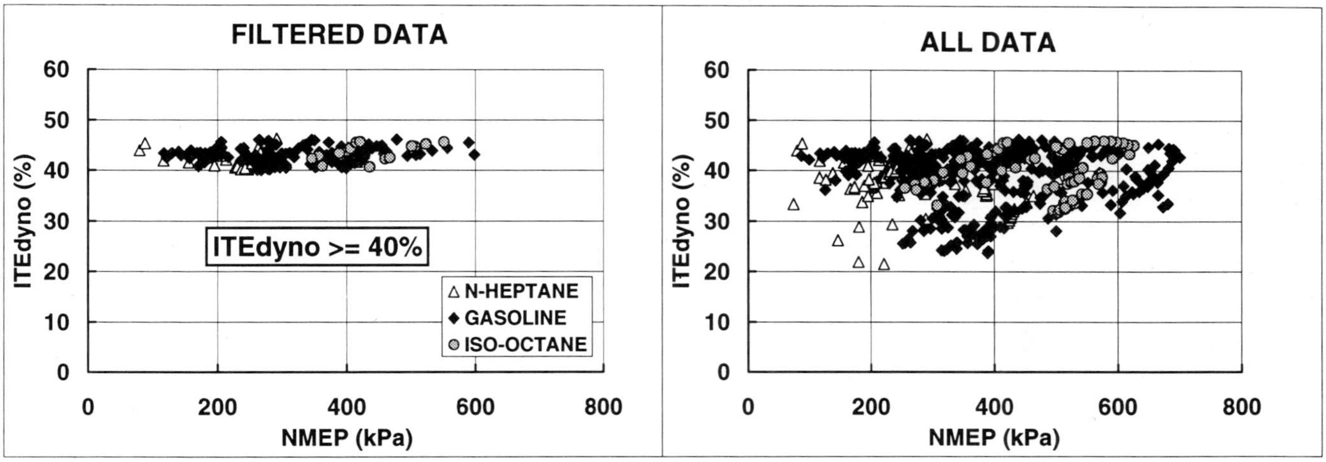

Figure 5 – Dynamometer-indicated thermal efficiency vs NMEP – for filtered and unfiltered data.

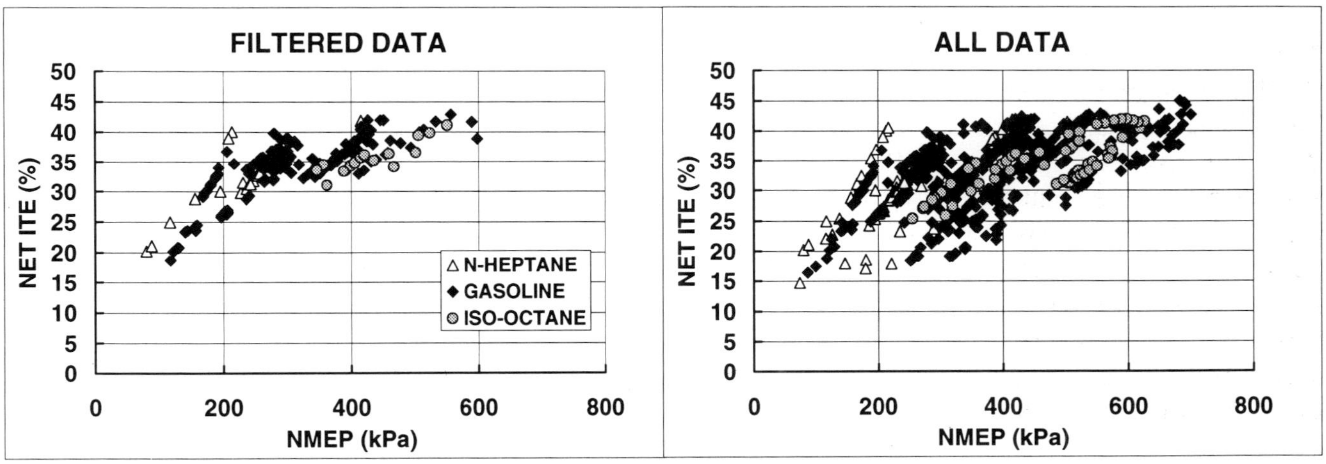

Figure 6 – Net indicated thermal efficiency (NITE) vs NMEP – for filtered and unfiltered data.

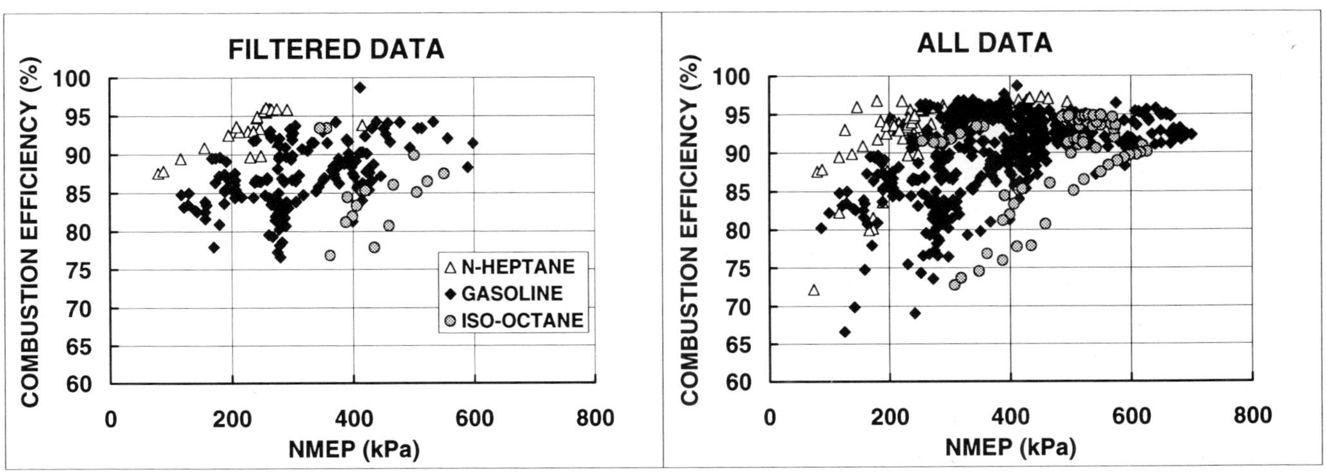

Figure 7 – Combustion Efficiency vs NMEP – for filtered and unfiltered data.

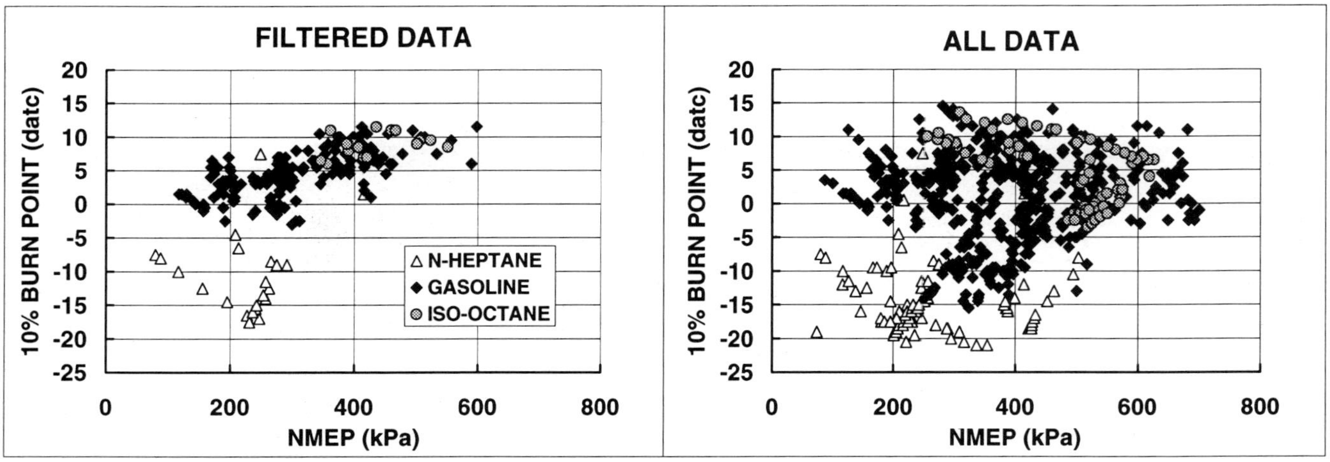

Figure 8 – Start of combustion (10% burn point) vs NMEP – for filtered and unfiltered data.

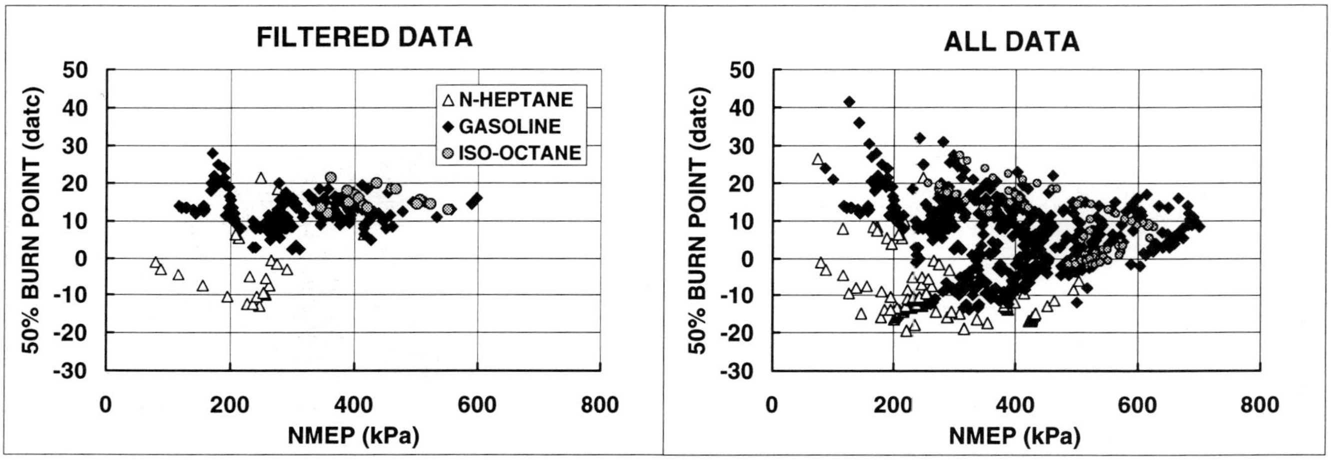

Figure 9 – Combustion phasing (50% burn point) vs NMEP – for filtered and unfiltered data.

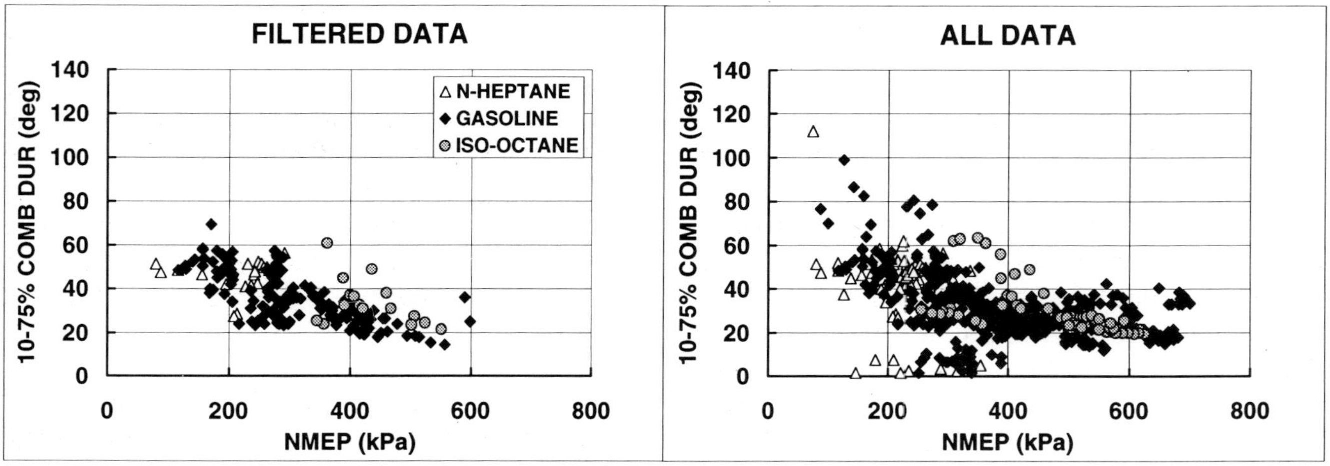

Figure 10 – 10-75% burn combustion duration vs NMEP – for filtered and unfiltered data.

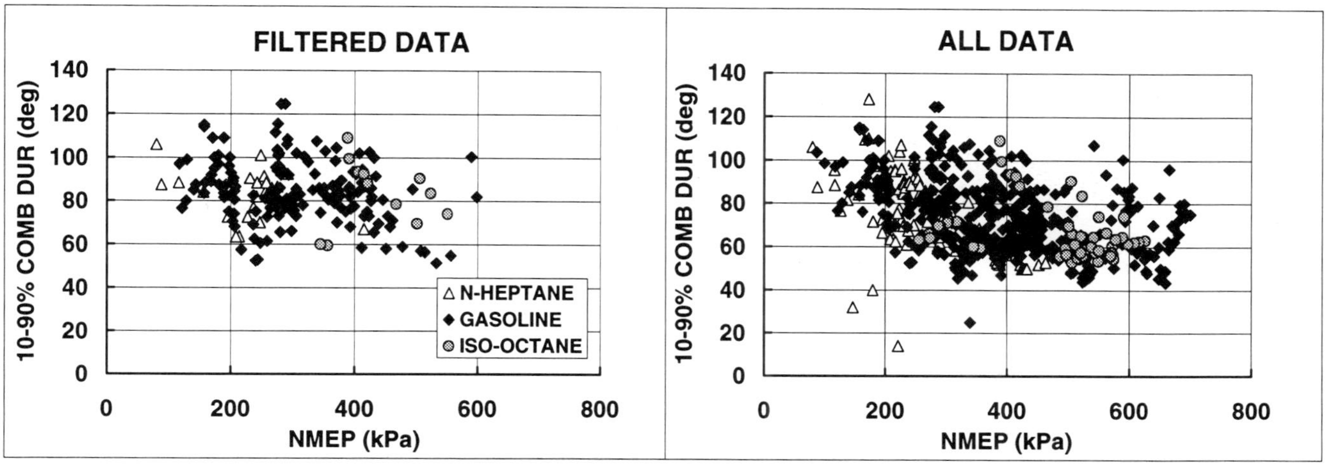

Figure 11 – 10-90% burn combustion duration vs NMEP – for filtered and unfiltered data.

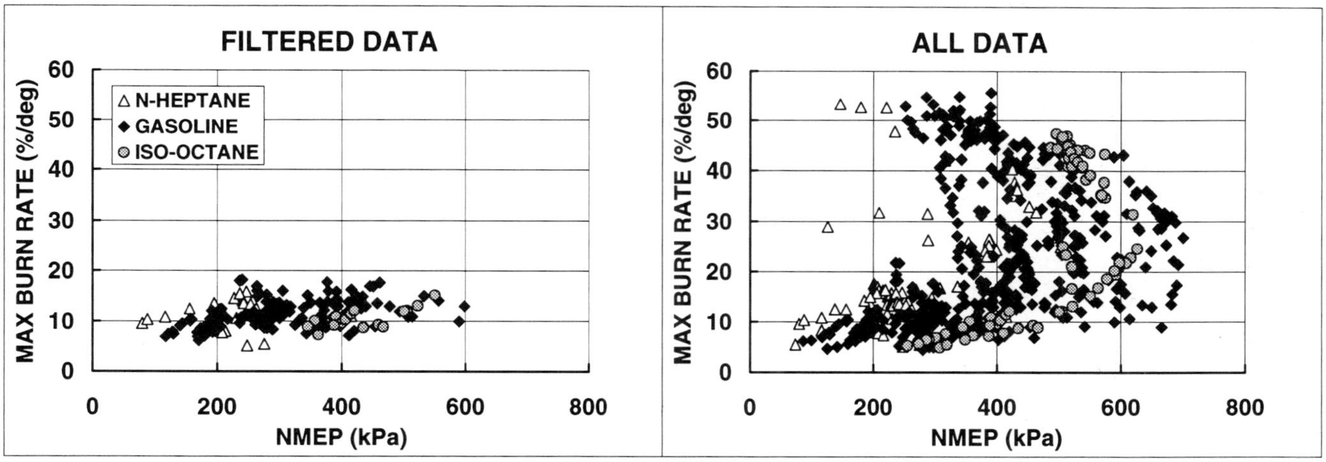

Figure 12 – Maximum burn rate vs NMEP – for filtered and unfiltered data.

Maximum-burn rate (Figure 12) and combustion phasing (Figure 9) combine to influence peak cylinder pressure, the location of peak cylinder pressure, and peak mean cylinder temperature (averaged over the cylinder contents) as shown in Figure's 13 through 15, respectively. For the filtered data, the range of peak pressure and location of peak pressure has been narrowed considerably. Maximum pressure has been lowed from about 12000 kPa to about 8000 kPa for the filtered case. Locations of peak pressure mostly fall between about 2 and 15 datc for the filtered case with gasoline.

Maximum mean cylinder temperature, given in Figure 15, is between 1000K and 1500K for the filtered data, consistent with the very low NOx emissions. Much higher values, approaching 2500K are observed for the unfiltered data. The high NOx emissions are associated with those high temperatures.

Maximum burn rate (Figure 12) has a direct effect on maximum rate of pressure rise, given in Figure 16. Recall that maximum pressure rise was constrained to be equal or less than 6000 kPa/ms (Table 2).

Maximum pressure rise is linked to combustion noise, and this can be seen in Figure 17 by comparison with Figure 16. The combustion noise meter filters the output of the cylinder pressure transducer and generates an output similar to the ringing intensity of a combustion knock analyzer.

Because the chosen noise meter output is a log scale, it is more appropriate to compare that value with the log of maximum pressure rise. In Figure 18, combustion noise meter values are shown to increase linearly with the log of maximum pressure rise.

Cyclic variability as indicated by the coefficient of variation (COV) of IMEP is generally low for HCCI combustion, Figure 19. For the filtered data, all values greater than 5% were excluded (Table 2).

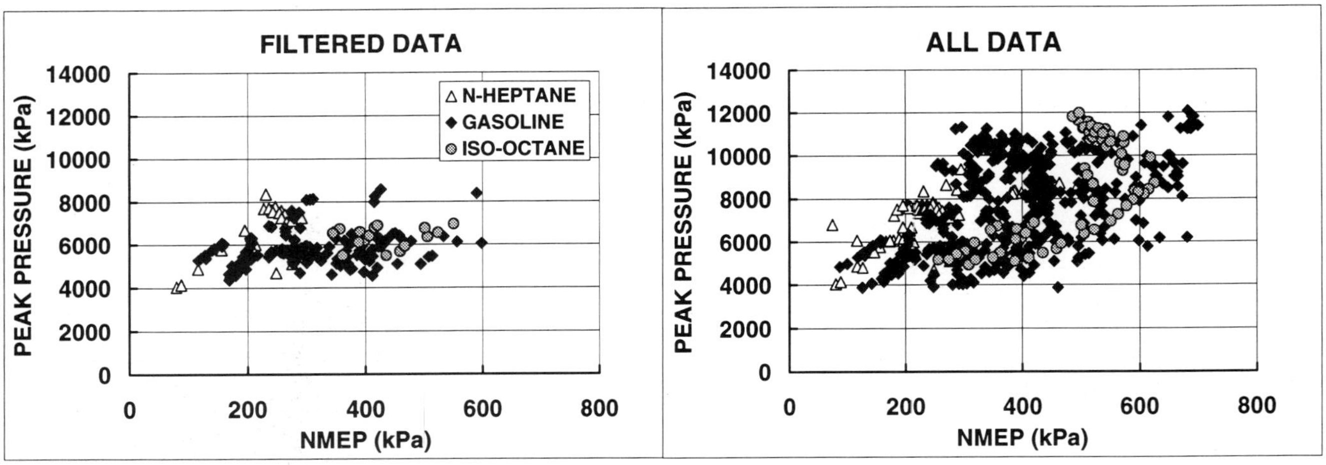

Figure 13 – Peak cylinder pressure vs NMEP – for filtered and unfiltered data.

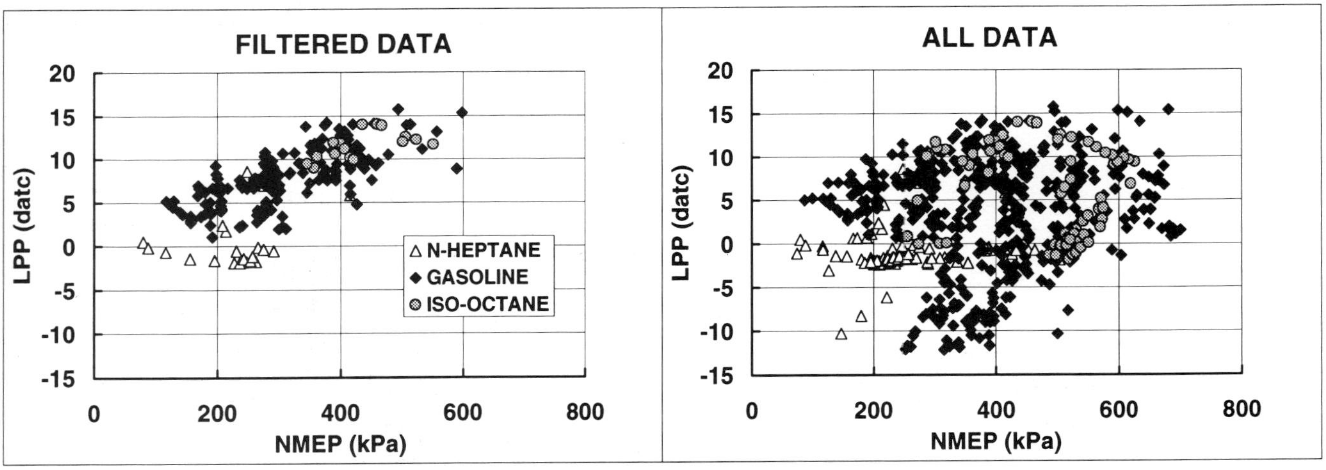

Figure 14 – Location of peak pressure vs NMEP – for filtered and unfiltered data.

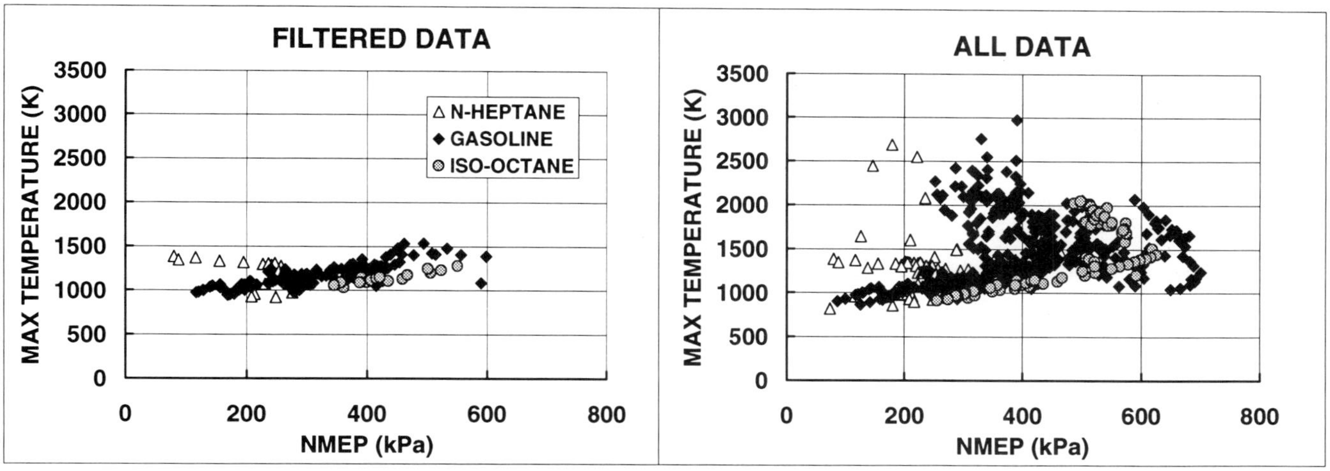

Figure 15 – Maximum cylinder temperature vs NMEP – for filtered and unfiltered data

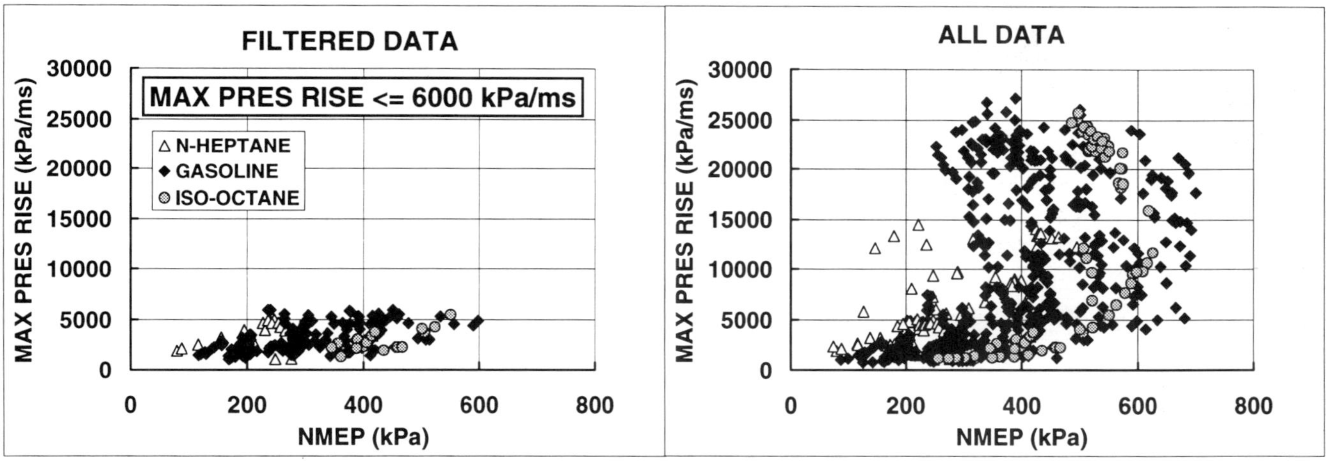

Figure 16 – Maximum pressure rise vs NMEP – for filtered and unfiltered data.

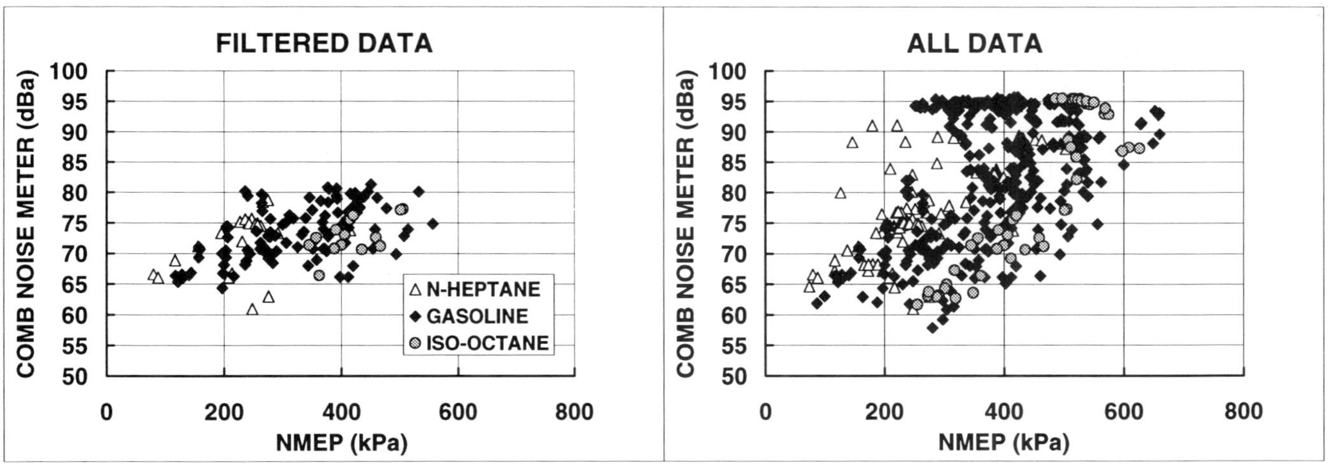

Figure 17 – Combustion noise meter output vs NMEP – for filtered and unfiltered data.

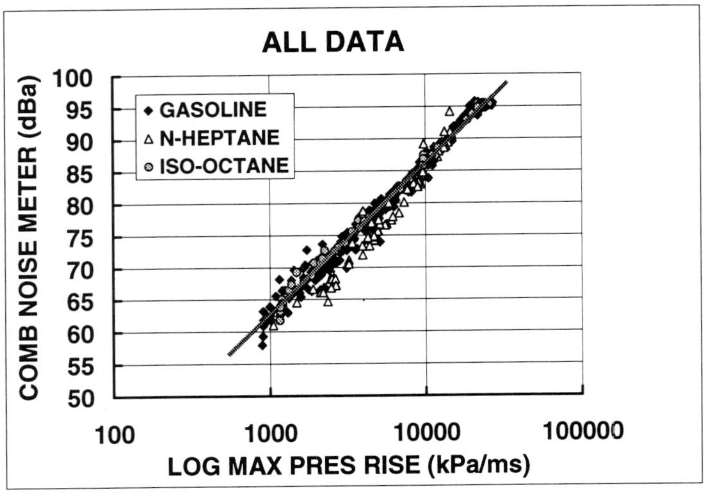

Figure 18 – Combustion noise meter output vs log maximum pressure rise – for all data.

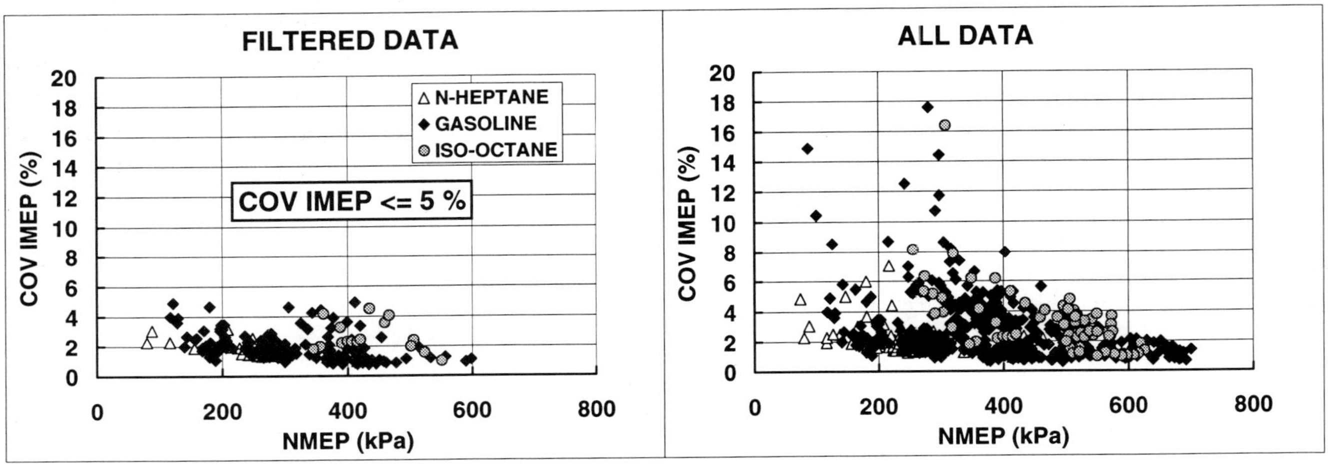

Figure 19 – Coefficient of variation in IMEP vs NMEP – for filtered and unfiltered data.

ACCEPTABLE OPERATING REGIONS FOR GASOLINE

The principal object of this work was to define the operating conditions where HCCI combustion is deemed acceptable. That was simply done by looking at the operating conditions for the filtered data with gasoline fuel. Thus, the acceptable operating regions for HCCI combustion with gasoline fuel are outlined in a series of figures to follow.

FUEL TYPE AND COMPRESSION RATIO - Of the fuels tested, multi-component gasoline (fuel A and B, Table 1) provided the greatest load range of operation, and for that reason was chosen over n-heptane or iso-octane as a potentially more useful fuel (for pre-mixed HCCI.) This enhanced load range is illustrated in Figure 20 where the RON of the fuels tested and the net MEP ranges are compared. The range of acceptable NMEP with gasoline is from 100 to 600 kPa. N-heptane, however, was limited to net MEP's below about 400 kPa, while iso-octane only operated acceptably above an NMEP of about 350 kPa.

While most of the data were taken at a compression ratio of 18.7:1, some also were taken at a lower compression ratio of 16.1:1. The effect of compression ratio on the load range is given in Figure 21. For the filtered data, lowering compression ratio has increased the minimum and maximum net MEP approximately 50 kPa.

INDEPENDENT OPERATING VARIABLES - The ranges of independent variable operating conditions for successful HCCI operation on gasoline are highlighted with free form, shaded areas overlaying the filtered HCCI data with gasolines A and B on Figures 22 through 25.

Fuel mass per cycle, Figure 22, varied from about 5 to 13 mg and increased in proportion to NMEP.

Manifold absolute pressure, given in Figure 23, varies from 100 kPa at the lowest NMEP to between 100 to almost 150 kPa at the highest NMEP.

Inlet temperatures, given in Figure 24, varied over a range from about 50 to 220°C. A trend is evident for the higher values at the lower NMEP's. At the time iso-octane was tested, there was insufficient air heater capacity to heat the inlet air above 100°C, and this temperature limitation contributed to that fuels inability to operate at lower NMEP's.

The range of EGR, Figure 25, varied from 0 to about 60% over the acceptable load range.

DEPENDENT OPERATING VARIABLES – Dependent variable operating conditions of interest are illustrated in Figures 26 and 27. Equivalence ratios, Figure 26, varied from slightly less than 0.2 to about 1.0, with the greater range favoring the higher NMEP's. The range of delivery ratio (volumetric efficiency), given in Figure 27, varies from about 30 to over 140%. The highest values occurred at the highest net MEP.

ENGINE SPEED AND SWIRL – The amount of data collected at speeds other than 1200 rpm or at increased swirl ratios were considered too limited to draw any conclusions as to either of those two variables. For the filtered data, all tests were at 1200 rpm and the lowest swirl level (both port throttle valves open.)

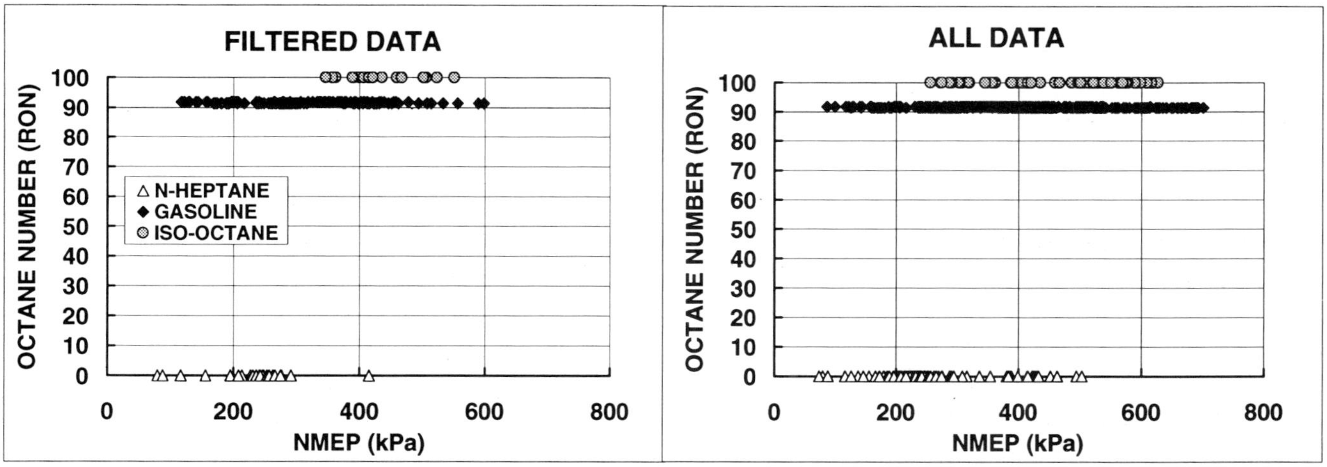

Figure 20 – Research octane number (RON) vs NMEP – for filtered and unfiltered data.

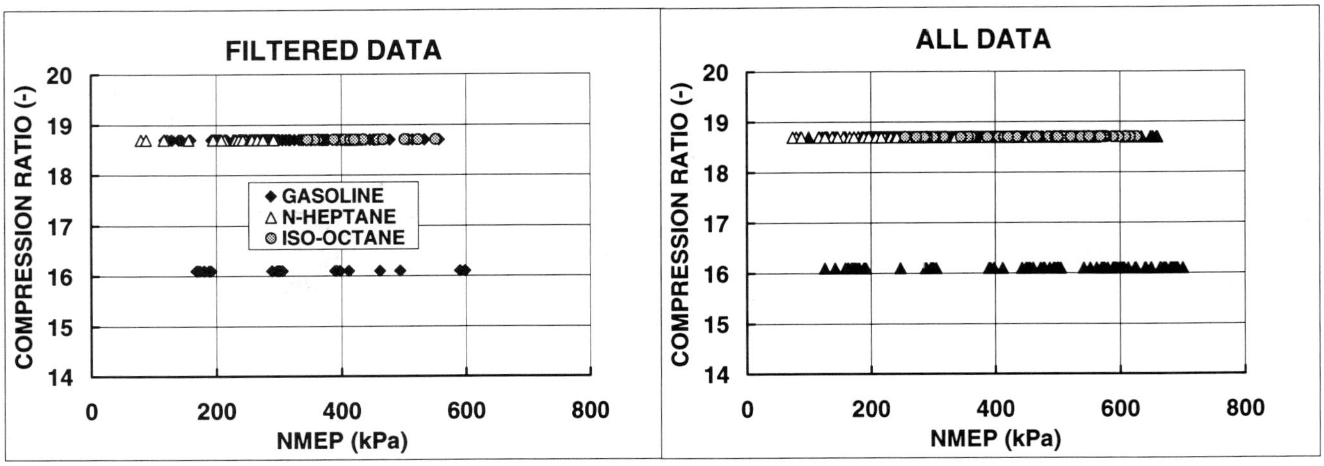

Figure 21 – Compression ratio vs NMEP – for filtered and unfiltered data.

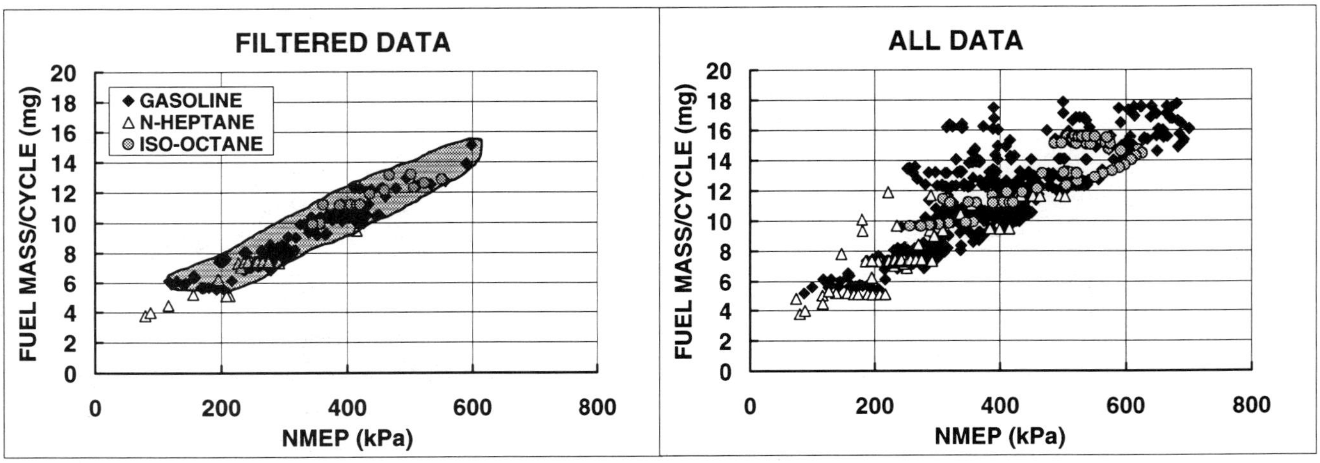

Figure 22 – Fuel mass per cycle vs NMEP – for filtered and unfiltered data.

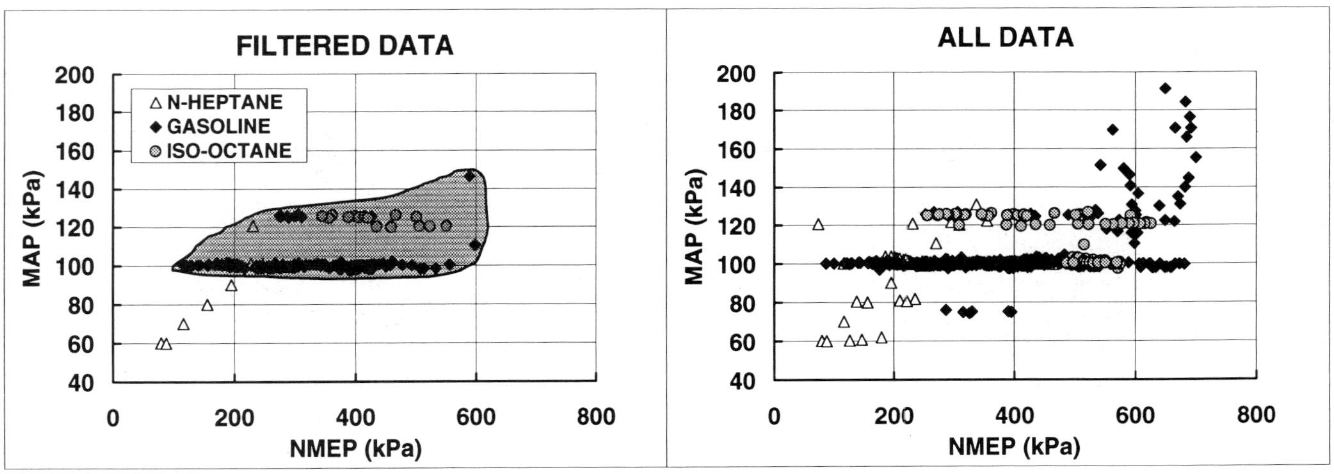

Figure 23 – Manifold absolute pressure (MAP) vs NMEP – for filtered and unfiltered data.

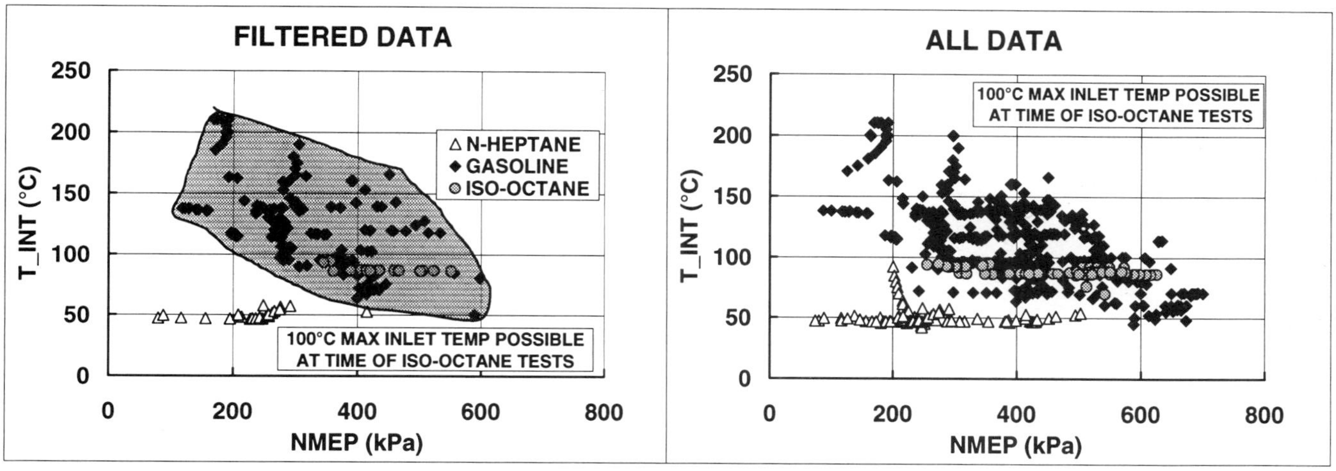

Figure 24 – Inlet temperature vs NMEP – for filtered and unfiltered data.

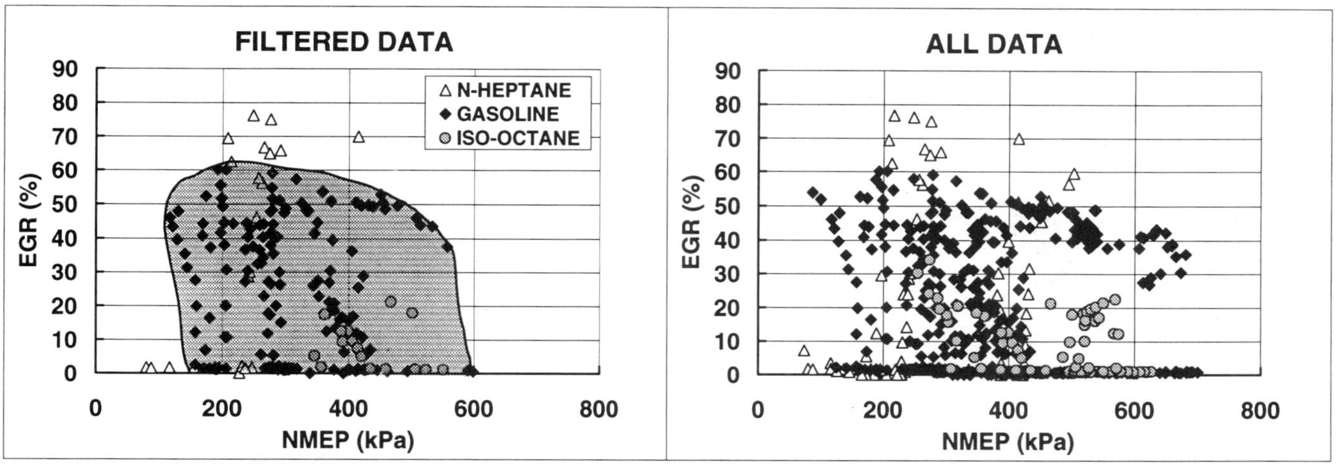

Figure 25 – Exhaust gas recirculation (EGR) vs NMEP – for filtered and unfiltered data.

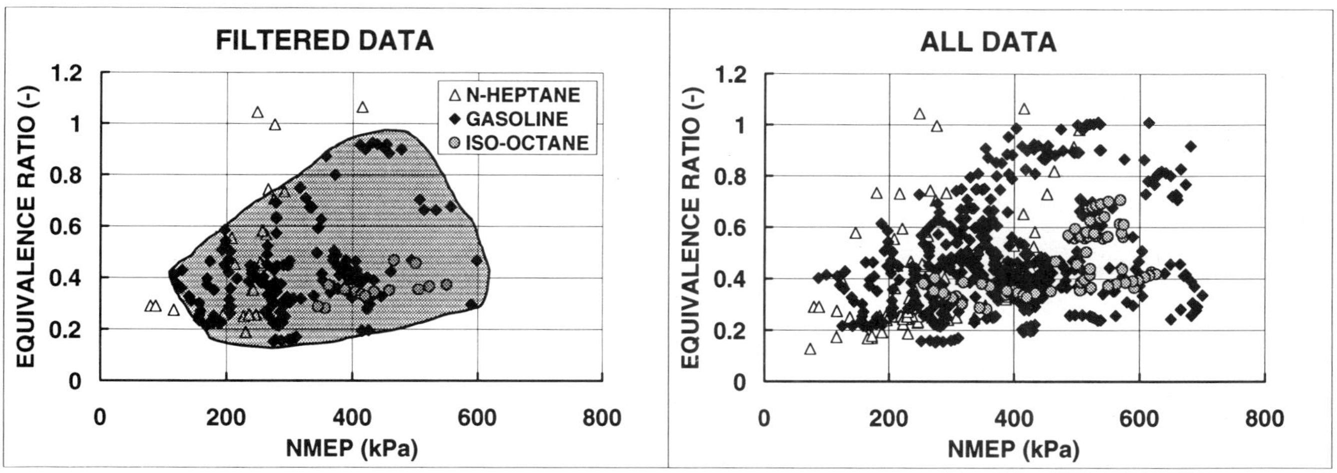

Figure 26 – Equivalence ratio vs NMEP – for filtered and unfiltered data.

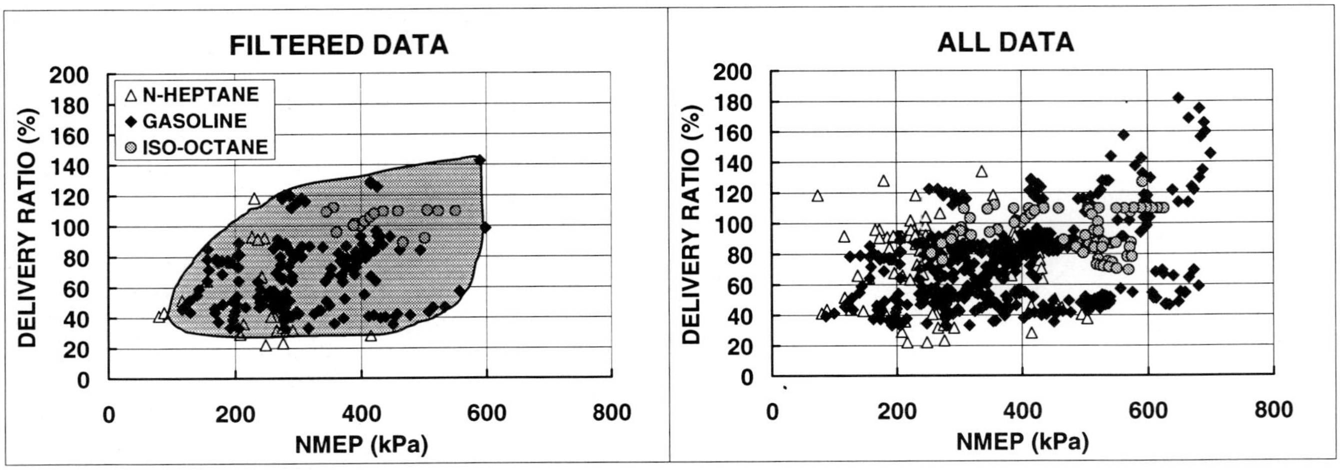

Figure 27 – Delivery ratio (volumetric efficiency) vs NMEP – for filtered and unfiltered data.

SUMMARY

A large data set of pre-mixed HCCI combustion taken from a single-cylinder diesel engine at two compression ratios (18.7:1 and 16.1:1) and with four fuels was analyzed for the operating range with acceptable HCCI combustion.

Acceptable HCCI combustion was defined as:

1. NOx emission index <= 1.0 g/kg
2. ITEdyno >= 40%
3. Max. Pres. Rise <= 6000 kPa/ms
4. COV of IMEP <= 5%

Multi-component gasoline fuels meet these criteria with the widest operating range (NMEP range from 100 to 600 kPa). N-heptane was limited to lower NMEP's (about 350-kPa maximum) and iso-octane to higher NMEP's (about 350-kPa minimum). The lower compression ratio raised both the minimum and maximum NMEP about 50 kPa.

The range of operating conditions for meeting the criteria included:

1. 5 to 13 mg fuel per cycle
2. 100 to 150 kPa MAP
3. 50 to 220°C inlet temperature
4. 0 to 60% EGR
5. 0.2 to 1.0 equivalence ratio
6. 30-140% volumetric efficiency

All of the data meeting the criteria were taken at 1200-rpm engine speed and with the lower swirl setting. Insufficient data were available at other speeds or swirl levels to include those variables in the operating range meeting the HCCI combustion criteria.

CONCLUSIONS

1. Acceptable HCCI combustion is limited to part load based on tests of pre-mixed gasoline fuels used in a diesel engine.
2. The load limitation is dictated by the dilute nature of the combustion process having peak mean temperatures of less than about 1500°K to prevent significant NOx formation.
3. Gasoline fuels provide a larger load range than the single-component fuels tested (n-hexane and iso-octane.)
4. Diesel fuel, assuming it could be vaporized to create the premixed mixture, would be expected to show a load limitation similar to that demonstrated by n-heptane.
5. When phased properly, thermal efficiency of HCCI combustion approaches that of a DI diesel, but falls short due to lower combustion efficiency and prolonged 10-90% combustion duration.
6. While the low NOx and smoke emissions of HCCI combustion is highly desirable, HC and CO emissions exceed those of a diesel and will have to be dealt with.

ACKNOWLEDGMENTS

Jim Cauley ran the HCCI experiments and collected all the data. His efforts are greatly appreciated.

REFERENCES

1. Siewert RM (1975) Staged Internal Combustion Engine and Methods of Operation. U.S. Patent 3,924,576, December 9, 1975.
2. Siewert RM (1975) Staged Internal Combustion Engine with Modulating Interstage Temperature Control. U.S. Patent 3,896,774, July 29, 1975.
3. Siewert RM (1976) Staged Internal Combustion Engine with Interstage Temperature Control. U.S. Patent 3,896,774, May 25, 1976.
4. Siewert RM, Turns SR (1975) The Staged Combustion Compound Engine (SCCE): Exhaust Emissions and Fuel Economy Potential. SAE Paper No. 750889.
5. Onishi S, Jo SH, Shoda K, Jo PD, Kato S (1979) Active Thermo-Atmospheric Combustion (ATAC)--A New Combustion Process for the Internal Combustion Engines. SAE Paper No. 790501.
6. Najt PM, Foster DE (1983) Compression-Ignited Homogeneous Charge Combustion. SAE Paper No. 830264.
7. Thring RH (1989) Homogeneous-Charge Compression-Ignition (HCCI) Engines. SAE Paper No. 892068.
8. Ryan TW, Callahan TJ (1996) Homogeneous Charge Compression Ignition of Diesel Fuel. SAE Paper No. 961160.
9. Gray AW, Ryan TW (1997) Homogeneous Charge Compression Ignition (HCCI) of Diesel Fuel. SAE Paper No. 971676.
10. Christensen M, Hultqvist A, Johansson B (1999) Demonstrating the Multi Fuel Capability of a Homogeneous Charge Compression Ignition Engine with Variable Compression Ratio. SAE Paper No. 1999-01-3679.
11. Takeda Y, Keiichi N, Keiichi N (1996) Emission Characteristics of Premixed Lean Diesel Combustion with Extremely Early Staged Fuel Injection. SAE Paper No. 961163.
12. Hashizume T, Miyamoto T, Akagawa H, Tsujimura K (1998) Combustion and Emission Characteristics of Multiple Stage Diesel Combustion. SAE Paper No. 980505.
13. Harada A, Shimazaki N, Sasaki S, Miyamoto T, Akagawa H, Tsujimura K (1998) The Effects of Mixture Formation on Premixed Lean Diesel Combustion. SAE Paper No. 980533.
14. Akagawa H, Miyamoto T, Harada A, Sasaki S, Shimazaki N, Hashizume T, Tsujimura K (1999) Approaches to Solve Problems of the Premixed Lean Diesel Combustion. SAE Paper No. 1999-01-0183.
15. Iwabuchi Y, Kawai K, Shoji T, Takeda Y (1999) Trial of New Concept Diesel Combustion System – Premixed Compression-Ignited Combustion –. SAE Paper 1999-01-0185.
16. Kong S-C, Marriott CD, Reitz RD, Christensen M (2001) Modeling and Experiments of HCCI Engine Combustion Using Detailed Chemical Kinetics with Multidimensional CFD. SAE Paper No. 2001-01-1026.
17. Duret P, Douaud A (2001) New Combustion Process: Homogeneous Diesel and Auto-Ignition Gasoline towards Zero Emission with Major Engine and Fuel Technology Challenges. Advanced Propulsion & Emission Technology, Vol. 18, Proceedings of the 2001 Global Powertrain Congress, June 5-7, 2001, Detroit.
18. Krieger RB, Siewert RM, Pinson JA, Gallopoulos NE, Hilden DL, Monroe DR, Rask RB, Solomon ASP, Zima P (1997) Diesel Engines: One Option to Power Future Personal Transportation Vehicles. SAE Paper 972683.
19. Kimura W, Aoki O, Ogawa H, Muranaka S, Enomoto Y (1999) New Combustion Concept for Ultra-Clean and High-Efficiency Small DI Diesel Engines. SAE Paper No. 1999-01-3681.
20. Kimura S, Aoki O, Kitahara Y, Aiyoshizawa E (2001) Ultra-Clean Combustion Technology Combining a Low-Temperature and Premixed Combustion Concept for Meeting Future Emission Standards," SAE Paper No. 2001-01-0200.

A New Generation of Engine Combustion Processes for the Future?
© P. Duret (Editor) and Editions Technip, Paris, 2001, pp. 25 - 34
27 rue Ginoux, 75015 Paris

VARIABLE VALVE ACTUATION FOR IN-CYLINDER CHARGE AND COMBUSTION CONTROL IN D.I. DIESEL ENGINES

Costantinos Vafidis, Antonio Piccinini, Laura Gianolio
Fiat Research Centre

ABSTRACT

The paper describes the design considerations and test results of the UNIAIR electro-hydraulic variable valve actuation system applied to the intake and exhaust valves of a 16-valve, 4-cylinder common rail DI Diesel engine for passenger car applications. The results indicate that the UNIAIR system exhibits accurate and flexible valve actuation characteristics under extreme in-cylinder pressure and oil temperature conditions. In turn, these valve actuation characteristics allow a continuous in-cylinder swirl modulation with minimal pumping losses, in-cylinder pressure and temperature control of the fresh air charge and, in particular, accurate stroke-to-stroke residual gas quantity control. When combined with external cooled EGR, the quantity, as well as temperature, of the residual gases can be modulated in a very wide range thus allowing the auto-ignition process of the in-cylinder charge to be efficiently controlled. The paper concludes that the charge control UNIAIR system, combined with the flexible "Multijet" fuel metering system can allow a realistic application of a modified combustion process in passenger car Diesel engines.

INTRODUCTION

Modern passenger car DI Diesel engines exhibit excellent power density, torque and driveability characteristics while maintaining at the same time excellent fuel consumption standards. This is due to the recent refinements of combustion control made possible by the advent of the common rail fuel injection system [1, 2] which are being further improved with the introduction of the second generation "Multijet" common rail system [3]. This concept allows multiple injections at very short intervals during an engine cycle, which shape the fuelling process in order to match the combustion requirements. Current DI Diesel engines comply with Euro 3 emissions limits while it has been shown that light vehicles equipped with more sophisticated Diesel engines [4] reach Euro 4 emission levels with excellent fuel consumption records [5]. However, reaching Euro 4 emission levels in higher segment vehicles equipped with DI Diesel engines is not trivial. The re-emerging particulate filter (DPF) technologies [6] are promising but durability and fuel consumption considerations make them unattractive. Similar arguments can be made for the Lean Nox Traps (LNT), which require very frequent regeneration thus seriously hindering the fuel economy benefit of the Diesel engine. Both technologies also suffer from increased costs and complex control, which often requires modification of the combustion process.

The alternative of reducing engine-out emissions through a radical modification of the combustion process is very attractive. Homogeneous Charge Compression Ignition (HCCI) as described in [7, 8 and 9] or, more realistically, Highly Diluted Low Temperature Combustion [10] processes are recently proposed and can potentially lead to extremely low emissions levels. It is, however, known that fine control by means of both, fuel injection and charge composition and motion characteristics is necessary in order to guarantee their applicability in the automotive non-stationary engine conditions, an issue that renders their implementation technically challenging.

The increased flexibility of fuel injection systems has been addressed during the last few years and its technical feasibility and positive effects on the combustion process and emissions have been discussed in [3]. This paper deals with the design and potential of a "charge control" device called UNIAIR, based on an electro-hydraulic variable valve actuation system already presented in [11] for gasoline engine applications, which is currently developed also for Diesel engine applications. The ambitious goal of this work is to combine the "fine fuel metering" of the Multijet injection system with a similar "fine gas metering" of the UNIAIR variable valve actuation system in order to provide the means necessary to control the modified, or non-conventional, combustion process. A further ambitious goal of this work is to maintain the high performance standards of the current D.I. Diesel engines which implies that the combustion system should be able to revert to the "conventional" one when the operating conditions call for it.

VARIABLE VALVE ACTUATION IN A DIESEL ENGINE

The role of variable valve actuation in a D.I. Diesel engine aiming to operate in a non-conventional combustion mode is that of controlling in-cylinder charge motion, composition and thermodynamic characteristics. When considering a modern, high specific power, Common Rail D.I. Diesel engine, the combustion system architecture should include four valves per cylinder. A fully flexible variable valve actuation system should ideally be able to guarantee independent intake valve actuation in order to allow the variation of the in-cylinder swirl level according to engine speed and load, while the actuation of the exhaust valves can be in tandem. An additional desirable feature of the variable valve actuation system would be its capability of actuating the intake and exhaust valves more than once during the engine cycle in order to modify the gas exchange process, and consequently the in-cylinder quantity of residual gases, according to engine speed and load. Further to the above, the in-cylinder swirl and residual gas control should be respectively independent and, most importantly, dynamic on a cycle-to-cycle and cylinder-to-cylinder basis.

The variable valve actuation concept suitable for a Diesel engine application may be mechanical, with limited flexibility and no dynamic cylinder-to-cylinder control, or camless hydraulic [12, 13] with full flexibility but possible penalties in terms of power consumption. The camless, electromagnetic system, e.g. [14], despite its flexibility, cannot be considered for such an application due to the incompatibility of its valve rest position at mid-lift and the small clearance of a Diesel combustion chamber at TDC. The intermediate solution represented by the UNIAIR electro-hydraulic system [11], characterised by adequate flexibility, low power consumption full compatibility with the D.I. Diesel engine architecture and intrinsic fail-safe features, has been preferred for the specific application.

THE UNIAIR VALVE ACTUATION SYSTEM

The UNIAIR variable valve actuation system is based on the so-called electro-hydraulic lost motion principle, which has been originally developed at Fiat Research Centre for throttle-less gasoline engine operation.

An illustration of the application of the UNIAIR system to a four-valve per cylinder DI Diesel engine is shown in Figure 1. The engine features two separate intake ducts and a siamese exhaust duct. One independent actuator each actuates the intake valves, featuring a tappet piston in contact with the cam, a high-pressure oil chamber and a second piston in contact with the valve stem. This second

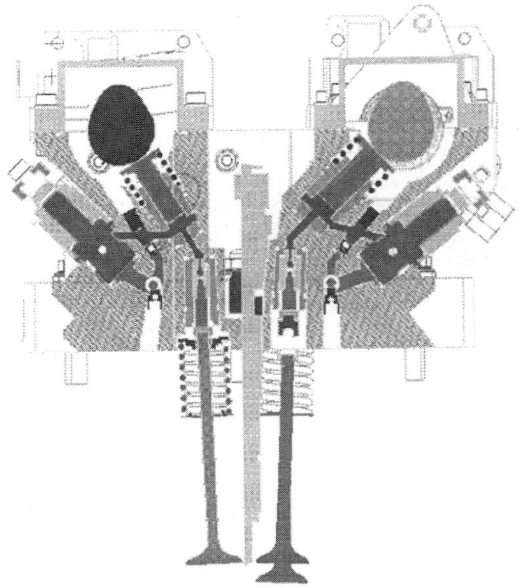

Fig. 1 The UNIAIR Actuator applied to a D.I. Diesel engine architecture.

piston incorporates the hydraulic brake that, during valve closing, ensures valve landing speed of less than 0.1 m/s. The intermediate oil chamber is connected via a channel to a specifically developed, normally open, solenoid valve which, when open, discharges the oil into a low-pressure channel incorporating a small number of spring-loaded accumulators. It is noted here that the oil used for the actuator is the engine lubrication oil itself. When the solenoid valve is energised (i.e. closed) the valve piston assembly follows the motion dictated by the cam profile. When early valve closing is required, the solenoid valve is de-energised (opens) and puts in communication the oil chamber with the accumulators. At this point the valve motion is de-coupled from the cam motion and closes under the force of the corresponding valve spring pushing oil through the solenoid valve and into the accumulator. As mentioned above, during the last phase of valve closure the hydraulic brake intervenes and softens the valve landing at the valve seat. The accumulators are primarily used in order to guarantee the immediate re-filling of the oil chamber when an immediate valve re-actuation is required. It is clear that, apart from the early valve closure, this system allows late valve opening if the solenoid valve is energised after the cam has started pushing the tappet piston. A single actuator incorporating one tappet piston operates each pair of the exhaust valves and one hydraulic brake assembly, which is in contact with the stems of the two exhaust valves by means of a bridge. The hydraulic circuit of the UNIAIR actuation system, as implemented to a four valve per cylinder D.I. Diesel Engine, is illustrated in Figure 2 illustrating also the relative position of the Common Rail fuel injector.

The above described system allows, therefore, the independent actuation of each intake valve of every single cylinder and the tandem actuation of each pair of the exhaust valves allowing also double actuations during an engine cycle.

Experimental evaluation of valve actuation

The above described UNIAIR valve actuation system has been designed using specifically developed hydraulic simulation codes which take into account the oil properties at various temperature and pressure levels (ranging between –30 and +140°C and up to 150 bar, respectively) and with various levels of air content, as well as the detailed geometric characteristics of the hydraulic lines and volumes.

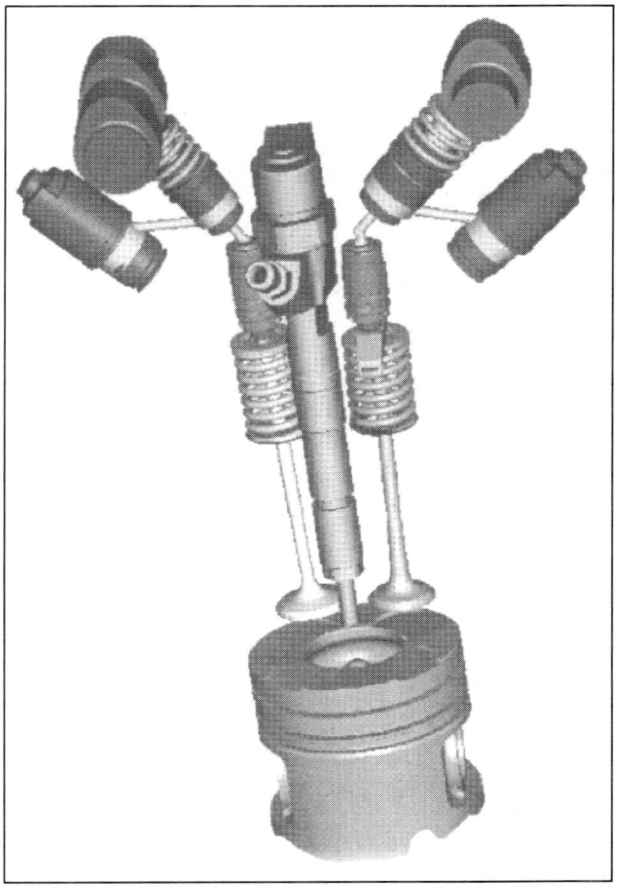

Fig. 2 Hydraulic circuit of the UNIAIR actuation system

The experimental verification of the functional characteristics of the actuator prototypes is performed in specific valvetrain evaluation test benches where the camshaft is motored at the desired speed, the oil and metal temperature is conditioned at the desired levels and the valve motion is monitored through high resolution and high frequency response, laser-based, non-contact displacement sensors pointing at the valve heads. The UNIAIR valve actuation system is in this case driven by its specific electronic control unit, which synchronises the valve actuation events with camshaft speed and phase.

An example of such measurements, which also illustrates the possible valve lift schemes that can be obtained with this actuator architecture, is shown in Figure 3 below.

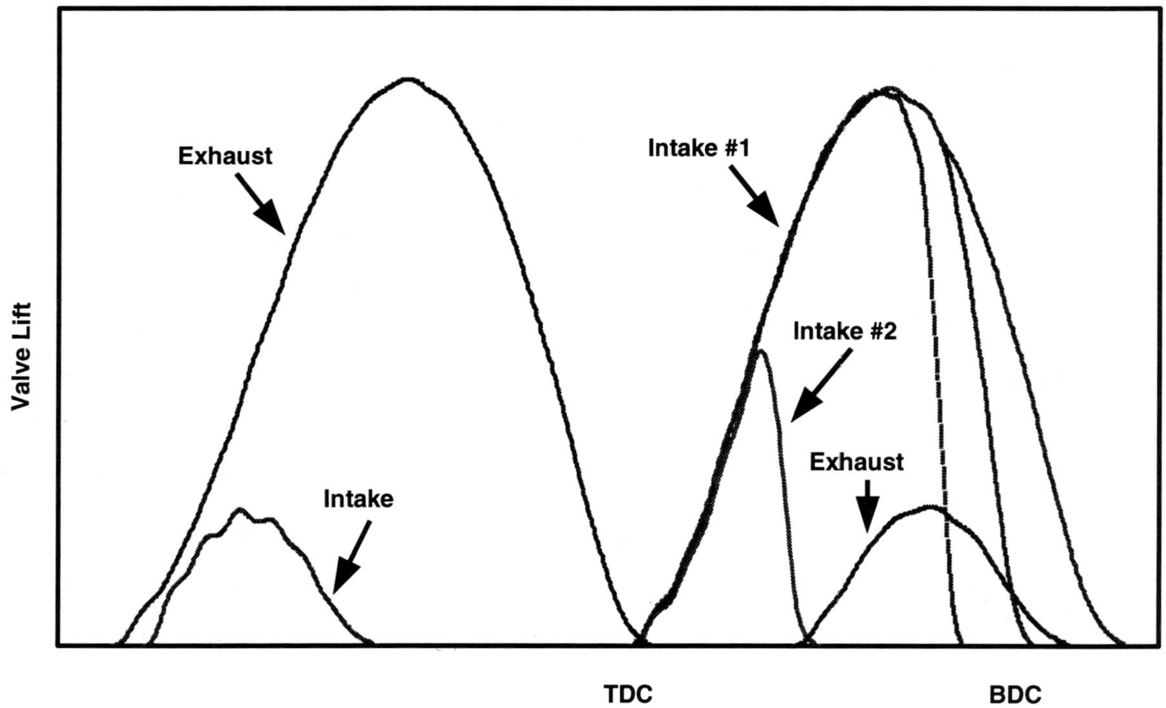

Fig. 3 Valve lift test results for various possible valve actuation schemes

Management of the gas exchange process

The flexibility of the valve control system described above applied to such a DI Diesel engine allows, in the first instance, the early closure of the intake valve(s), at around BDC during engine cranking. This increases the effective compression ratio by approximately one unit, thus allowing the engine geometric compression ratio to be reduced to levels of the order of 16:1 or less. The advantage of this reduction is dual since it allows higher torque for a given maximum in-cylinder pressure but also more excess air and lower in-cylinder pressure and temperature at part loads with benefits on NOx and PM emissions.

The independent actuation of the two intake valves effectively allows the continuous variation of the swirl ratio; the possibility of partially opening one valve for a very brief duration also compensates the otherwise increased pumping losses without reducing significantly the swirl level. It has been demonstrated that such strategy radically improves the soot-NOx trade-off particularly at low engine speeds without significant effect on fuel consumption.

Swirl Modulation

This procedure for in-cylinder swirl control through intake valve differentiated lift control has been analytically studied using a thermodynamic wave-action engine simulation model in conjunction with swirl and discharge coefficient measurements of the specific intake duct architecture for a variety of possible lift combinations. Some results are summarised in Figure 4 below in terms of calculated angular momentum of the in-cylinder charge at inlet valve closure for four valve actuation strategies.

The valve actuation mode A corresponds to both valves opening at full lift while mode B correspond to one only valve opening. Valve actuation modes C and D regard partial opening of one valve as illustrated in Figure 5 below.

As shown in Figure 4, the resulting in-cylinder swirl in the actuation mode C is very similar to that of mode B, while mode D produces high swirl levels with small penalty in terms of pumping work as illustrated in Figure 6 below, for various engine speeds and loads.

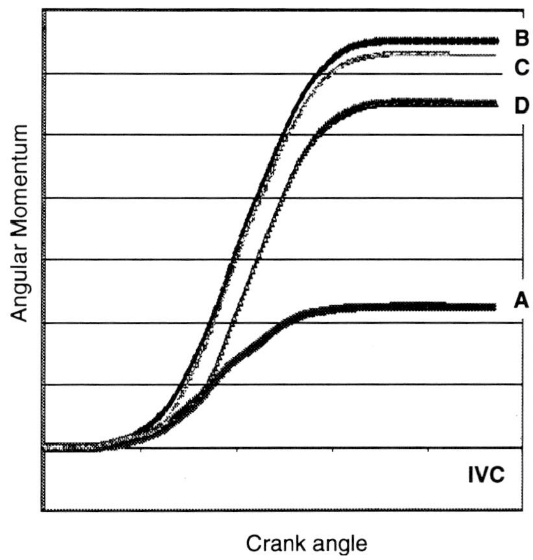

Fig. 4. In-cylinder angular momentum for various intake valve actuations

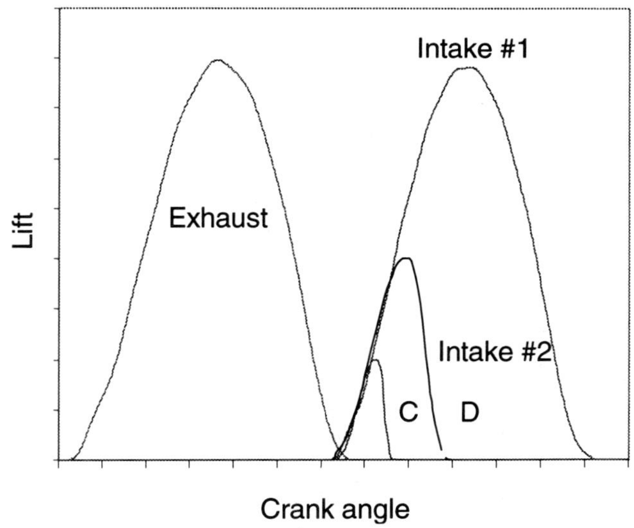

Fig. 5 Differentiated intake valve actuation modes.

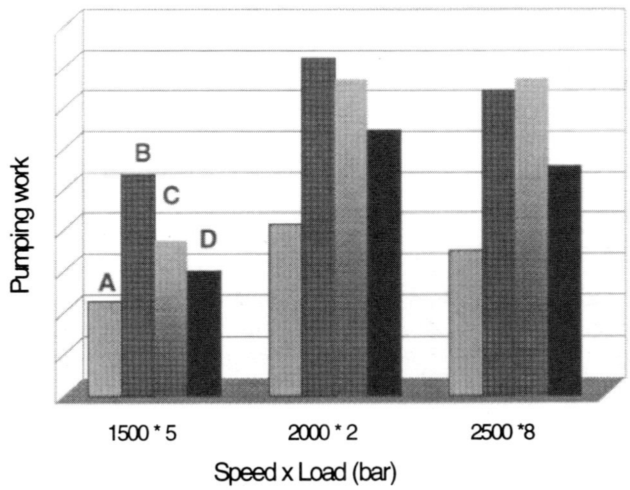

Fig. 6 Pumping work for the various valve actuation modes.

Residual Gas Control

The double actuation of the intake and/or exhaust valves, as shown in Figure 3, allows the modification of the gas exchange process in such a manner that the quantity of residual combustion gases (internal EGR) can be modulated with high precision and on a stroke-by-stroke basis to levels higher than 60%. An example of calculated instantaneous mass flow rates through the engine intake and exhaust ports is shown in Figure 7.

The engine operating point is 1000 rpm at load of 2 bar bmep and the valve actuation corresponds to full lift of both intake and exhaust valves and full secondary lifts of both valves during the exhaust and intake phases respectively as shown in Figure 3. The results of the calculations indicate back flow of burnt gases into the intake port during the exhaust phase and their subsequent re-breathing during the intake phase accompanied by exhaust gas re-breathing during the last phases of the intake stroke through the exhaust port. The resulting overall percentage of burnt gases trapped into the cylinder at the start of the compression stroke (internal EGR) exceeds 60%.

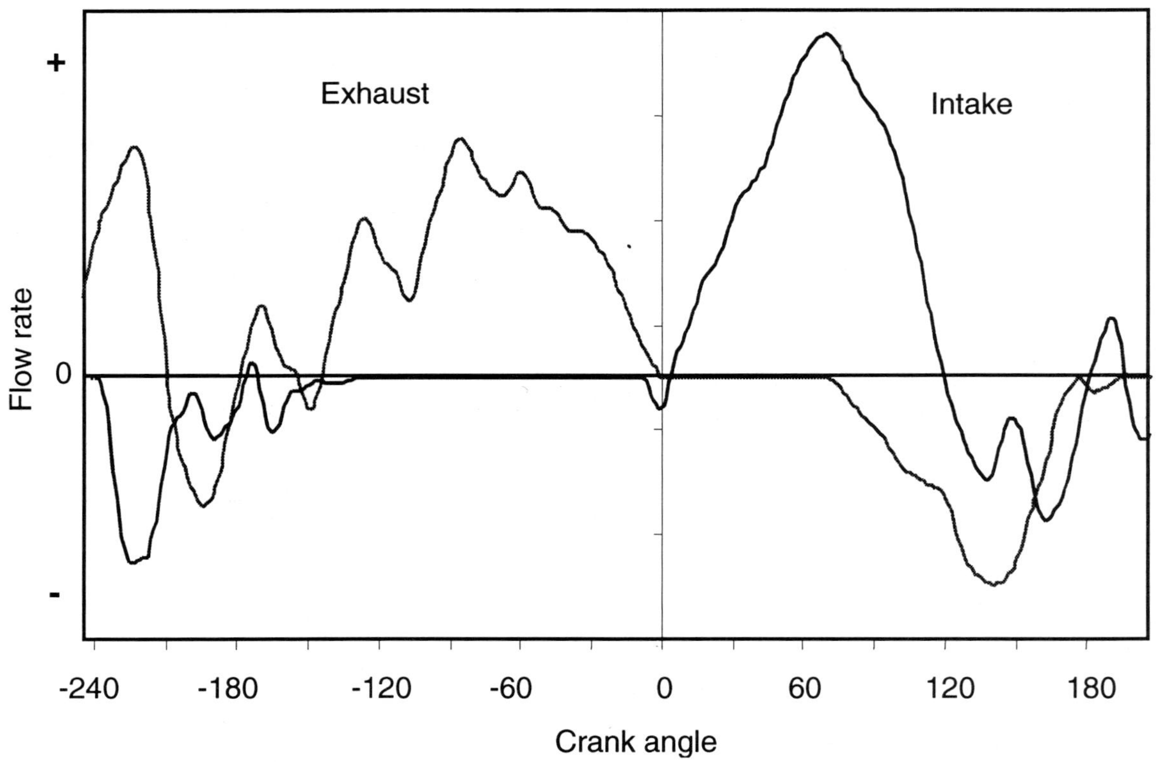

Fig. 7 Air and burnt gas flow rate through the intake and exhaust ports at 1500 rpm and 2 bar bmep.

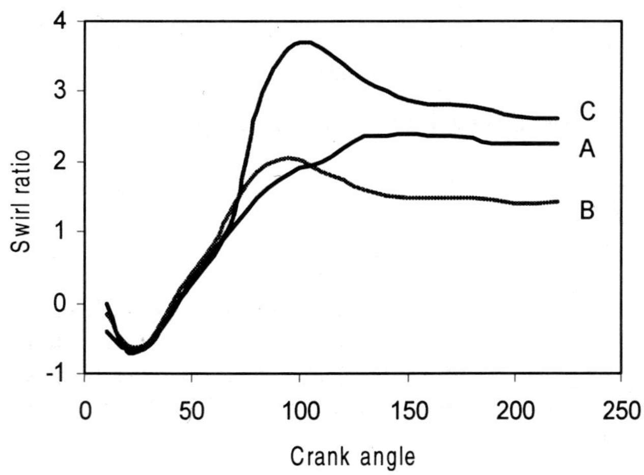

Fig. 8 In-cylinder swirl ratio variation as a function of crank angle (1000 rpm, 2 bar bmep).

The disadvantage of this procedure of trapping residual gases into the cylinder through re-activation of the exhaust valves during the intake stroke is a marked reduction of the in-cylinder swirl level due to the introduction of a mass of burnt gases with no or low (or even negative) angular momentum with respect to the cylinder axis, depending on the port layout. For the particular engine architecture a detailed CFD analysis was performed which indicated a potentially drastic reduction of the in-cylinder swirl level, as shown in Figure 8. Case A refers to the standard valve actuation without valve re-opening for residual gas control. Case B corresponds to the one described earlier (Fig. 7) where the valves are re-actuated in order to trap residual gases and highlights the significant decrease of the swirl level.

Case C refers to a correcting action, which recovers the in-cylinder swirl level while guaranteeing the process of burnt gas re-breathing. In this case one of the intake valves is opened only partially during the intake stroke, as described in earlier sections, thus enhancing the in-cylinder swirl generation during intake. The subsequent reduction of swirl during the exhaust gas re-breathing process through the exhaust valves, although reducing swirl, allows the recovery, or even enhancement, of the baseline swirl level.

The above examples have illustrated that the UNIAIR variable valve actuation system allows in-cylinder swirl and residual gas control with a high degree of flexibility. This capability of controlling the gas exchange process opens new horizons for combustion control, in particular the so-called "highly diluted" or "low temperature" combustion, which can be obtained in the presence of extreme levels of residual gases, [10]. Such combustion process, taking place at temperatures below 2000°K nearly eliminates NOx formation and, when below 1400°K it can eliminate soot. The necessary delicate control of mixture dilution and combustion rate can be obtained through the UNIAIR system.

Charge Temperature Control

The modified gas exchange process described above, although very efficient in controlling dynamically the amount of residual gases in the combustion chamber, does not provide means of controlling their temperature. It is known that efficient NOx emissions reduction requires low temperature EGR, which is currently obtained through the external EGR cooler.

The combustion system presented here does not preclude the use of external, cooled, EGR. In fact, it has been shown that the internal and external EGR can be used simultaneously with, nearly, synergic result reaching levels of residual gases trapped into the combustion chamber in excess of 80%. In this case the charge temperature can be controlled by dosing the appropriate amounts of external (cold) and internal (hot) EGR. It is clear that, in this case, the dynamic control of EGR is not feasible and that the charge temperature cannot be reduced significantly. The UNIAIR system also allows an additional, although less efficient, means of cooling down the in-cylinder charge at very low speeds and loads through early intake valve closing followed by charge expansion and re-compression. The same process also permits the variation of the effective compression ratio, which can control the cylinder pressure near the ignition angle.

From the above it is concluded that the UNIAIR valve control system allows in principle the variation of the thermodynamic conditions of the charge in the combustion chamber in terms of composition, pressure and temperature. This, combined with the capability of modulating the in-cylinder flow field, should provide an adequate number of controlling parameters in order to ensure combustion of a highly diluted charge through the control of the auto-ignition delay.

A most important feature of the proposed system is the fact that it maintains the main combustion system characteristics of a conventional D.I. Diesel engine. This allows the operation in a conventional combustion mode outside the emissions zone, thus guaranteeing the expected high engine performance. In fact, the proposed low geometric compression ratio and the efficient handling of the "post-charging" mechanism at low speeds and maximum load [15] should further enhance the overall engine performance. Figure 9 illustrates a possible valve actuation strategy that could be used in the entire engine map.

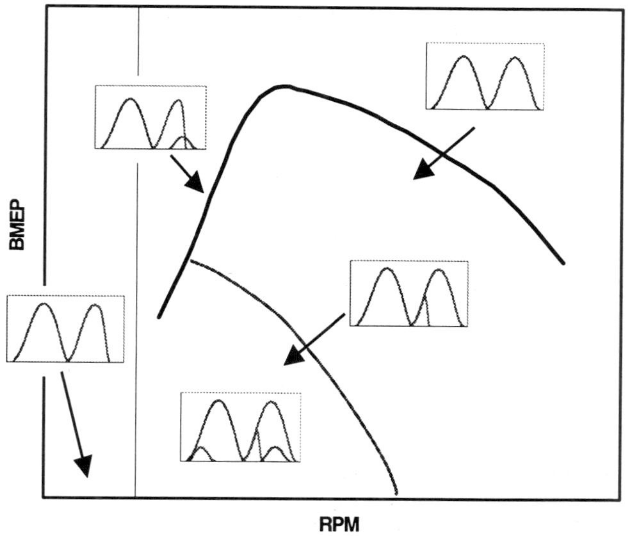

Fig. 9 Possible valve actuation strategy.

FUELLING CONTROL

Further to the charge composition and motion control through the UNIAIR valve actuation system, the fuelling process must be flexible enough to guarantee the optimal fuel distribution in the combustion chamber.

The Multiple Injection (Multijet, [3]) concept is based on the exploitation of the intrinsic flexibility of the common rail fuel injection system. In principle the overall injection process can be split in a sequence of several elementary injections with a minimum dwell time between two consecutive injections of the order of 500 microseconds.

The Multijet system currently allows up to five consecutive injections within a complete combustion cycle. At present the injection strategy may include a Pilot injection, a Pre injection, a Main injection, an

After injection and a Post injection. The Pilot injection increases the cylinder pressure and temperature and reduces the ignition delay of the Main injection, with a consequent reduction of the combustion noise, albeit with an increase of CO and Soot due to incomplete oxidation of the burnt fuel. The Pre injection acts on the combustion process by imposing a more gradual heat release in its first stages. This leads to a simultaneous reduction of combustion noise, Soot and CO emissions. Moreover the sequence of Pilot, Pre and Main injections has a beneficial effect on combustion control in cold operation improving the misfiring versus noise trade-off and thus allowing a compression ratio reduction. The possibility of using a very small dwell time between Pre and Main injections provides an improvement in Soot versus noise trade-off leading to the possibility of introducing higher EGR rates with a consequent NOx reduction. If the Main injection is followed by a close-coupled After injection, an increase in temperature during the diffusive combustion phase can be obtained allowing the post oxidation of the Soot produced in the earlier stages of combustion and resulting in its overall reduction. This reduction in Soot emissions allows a further increase of the EGR rate leading to an overall reduction of NOx and PM.

In current Euro4-oriented applications, the After injection, in conjunction with a retarded Post injection, allows the implementation of DPF regeneration strategies without fuel additives. A dedicated management of After and Post injection enables the increase of the exhaust gas temperature but also the increase of the catalyst outlet temperature through the addition of unburned hydrocarbons, thus creating conditions favourable for oxygen-based DPF regeneration.

The “conventional” emissions control techniques described above can always be implemented in the proposed concept of D.I. Diesel engine in operating zones where the modified, Low Temperature or HCCI, combustion processes are not feasible. The flexibility of the Multijet system, however, can also be exploited in the modified combustion regimes where its capability of handling small fuel packages at short intervals may become crucial for fuel mixture homogenisation. Optimistically, it can be envisaged that early multiple injections, combined with the vigorous in-cylinder gas motion and the high charge temperature due to the internal EGR, can lead to a fuel/air mixture homogeneity suitable for the HCCI process. A more realistic view may be that of using the Multijet’s capability of performing Pre injection and split Main injections in order to control the auto-ignition delay and combustion phase in the presence of extreme EGR levels aiming to implement the Low Temperature Combustion mode.

CONCLUSION

This paper presented the design features and functional characteristics of the UNIAIR variable valve actuation system implemented to the intake and exhaust valves of a Common Rail D.I. Diesel engine.

It was shown that the consequent flexible valve actuation allows accurate dynamic cylinder-to-cylinder and cycle-to-cycle modulation of the in-cylinder swirl level, quantity of residual gases, charge pressure and, to a lesser extent, charge temperature. The control of these parameters, combined with the flexibility of the fuelling process provided by the Multijet injection system, opens new horizons for the implementation of a non-conventional combustion process in an automotive D.I. Diesel engine. Such combustion process, which can be described as “Homogeneous Charge Compression Ignition“ or, more realistically, “Highly Diluted Low Temperature Combustion”, should lead to near zero NOx and soot emissions in a certain low load zone of the engine operating range. Outside this zone, the proposed system ensures conventional engine operation.

REFERENCES

1. Stumpp, G. and Ricco, M. "Common Rail - An attractive fuel injection system for passenger car DI Diesel engines", SAE Paper 960870, 1996.

2. Rinolfi, R., Imarisio, R. and Buratti, R. "The potential of a new Common Rail Diesel engine injection system for the next generation of DI Diesel engines", 16th International Vienna Motor Symposium, Vienna, 1995, VDI-Vrlag Reihe 12, Nr 239.

3. Imarisio, R., Ricco, M., and Rossi Sebastiano, G.M. “Multiple Injection, a cost effective solution for emission reduction of Common Rail DI Diesel Engines“, Paper to be presented at Aachen Colloquium - Automobile and Engine Technology, Aachen, October 4-6, 2000.

4. Bernard, L., Occella, S. and Vafidis, C. “Potential of a new very small DI Diesel engine for passenger cars” Paper F98T134, FISITA World Automotive Congress, Paris, September 27 - Oct. 1 1998.

5. Bernard, L., Casalone, M. and Rossi Sebastiano G.M. "Achieving very high fuel economy and good performance with engine downsizing: the Ecobasic concept car", First Ricardo International Conference, Brighton, June 21-23, 2000.

6. Solvat, O., Marez, P. and Belot, G. "Passenger Car Serial Application of a Particulate Filter System on a Common Rail Direct Injection Diesel Engine", SAE Paper 2000-01-0473.

7. Pucher, G.R., Gardiner, D.P., Bardon, M.F. and Battista, V. "Alternative Combustion Systems for piston engines involving Homogeneous Charge Compression Ignition concepts- a review of studies using methanol, gasoline and diesel fuel", SAE Paper 962063.

8. Christensen, M., Johansson, B., Amnéus P. and Mauss F. "Supercharged Homogeneous Charge Compression Ignition", SAE Paper 980787.

9. Gray, A. W. B. and Ryan, T. W. "Homogeneous Charge Compression Ignition (HCCI) of Diesel Fuel", SAE Paper 971676.

10. Sasaki, S., Ito, T. and Iguchi, S. "Smoke-less Rich combustion by Low Temperature Oxidation in Diesel Engines" Aachener Kolloquium, Aachen, October 4-6 2000.

11. Bernard, L., Ferrari A., Vattaneo, F. and Vafidis, C. "Developments in Fiat's electro-hydraulic variable valve actuation system", Congress on Variable Valve Timing – Benefits of Variable Valve Timing to Reduce Fuel Consumption and Emission, Haus der Technik, Essen, March 28-29, 2000.

12. Buehrle, H., Clark, R., Gross, J., Long, R. and Nist, L. "Internal combustion engine valve operating mechanism", U.S. Patent No. 6,024,060, February 15, 2000.

13. No Author, "Daimler Benz Research in Germany: Electrohydraulic valves replace camshafts", Machine Design, News Off The Wire, June 18, 1998, p. 25.

14. Pischinger, M., Salber, W., Van der Staay, F., Baumgarten, H., Kemper, H. "Benefits of the electro-mechanical valve train in vehicle operation", SAE Paper 2000-01-1223

15. Vafidis, C. "The Application of an Electro-hydraulic VVA System on a Passenger Car C.R. Diesel Engine", Int. Symposium "The Future of Diesel Engine Technology for Passenger Cars", Porto Cervo, October 12-13, 2000.

A New Generation of Engine Combustion Processes for the Future?

27 rue Ginoux, 75015 Paris

IGNITION TIMING CONTROL AT TOYOTA "UNIBUS" COMBUSTION SYSTEM

Hiromichi Yanagihara
Toyota Motor Europe

ABSTRACT

Several approaches are being taken to develop new combustion systems for drastic reductions of nitrogen oxide and soot emissions from diesel engines, while remarkably increasing fuel economy. The most promising approach is one in which a proper combustion structure is provided with a premixed compression ignition system. The author et al. proposed a new system named UNIBUS (Uniform Bulky combustion System). The image of combustion was taken by shadow graph method, and an image intensifier was used to detect very weak photic signals in combustion. The level of NO_x is about 1/100 that of a conventional DI diesel engine. The mechanism of extremely low NO_x and the onset of ignition were discussed. The study revealed that a new combustion process can be established by promoting fuel dispersion before ignition. The pre-reaction prior to combustion was intensively discussed from the aspects of mixture formation and chemical reaction. The chemical reactions such as from low temperature oxidation to high temperature oxidation which occur due to the gas temperature changing in engine, were determined by analyzing the combustion images and the heat release patterns. Furthermore, in application of the concept to engines with the usual Common Rail injection system and hole type nozzle, the new idea of ignition timing controlling was introduced additionally in the original concept.

INTRODUCTION

For drastic reductions of pollutant emissions while remarkably improving fuel economy from diesel engines, several research works have been reported[1]-[8] . The most promising approach is HCCI (Homogeneous Charged Compression Ignition) engine. A number of studies on combustion with HCCI engine have been conducted and reported so far [9]-[11]. The author et al. propose a new system named UNIBUS (Uniform Bulky combustion System) which falls in the category of systems for diesel engine combustion with the homogeneous charged compression ignition. The great concern of HCCI engine is how to control the combustion rate. In this paper, the combustion system of UNIBUS and its ignition timing control are investigated from aspects of thermodynamics and chemical reaction phenomenon in combustion.

CONCEPT OF MIXTURE FORMATION

At the preliminary test, the higher responded injection system with piezo actuator was used to form homogeneous mixture. Figure 1 shows the piezo-actuated injection system with a pressure chamber for fuel accumulation (Accumulated Pressure Injection System). The both Piezo-actuator and accumulation chamber located

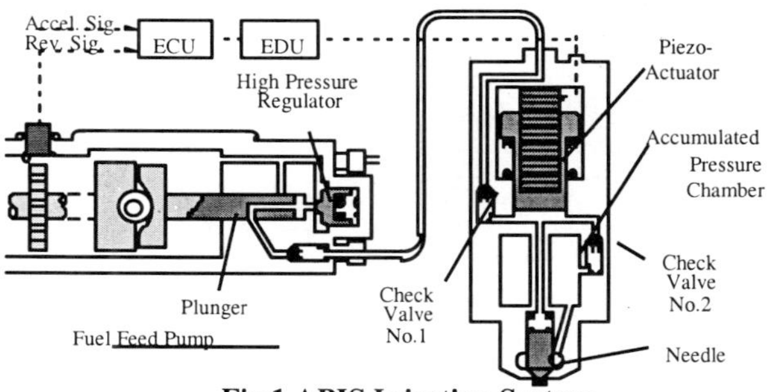

Fig.1 APIS Injection System

close to outlet of nozzle enabled the short injection duration . Figure 2 shows the bottom and side views of the spray image of the modified pintle type nozzle and the conventional multi-hole type nozzle under pressurized N2 gas. With the modified pintle nozzle, fuel spray spread in a hollow cone, in which fuel dispersion progressed to a great extent, but its penetration was suppressed, in comparison to the conventional multi-hole nozzle. The spray penetrated slow even immediately after injection, and never reached the cylinder wall. Figure 3 shows that injection rate by piezo injection system is compact. The initial rising rate of needle opening was very quick. Thus, the quick response injection system could inject fuel into a broad space without reaching the cylinder wall, since its injection rate decreased rapidly at onset of closing of nozzle needle. In this case, the injection pressure was set at 50Mpa. The pressure level was lower than that of conventional high pressure injection system. Obviously, for HCCI injection system, the time response seemed to be very important to make homogeneous mixture. By consideration with the experimental results, the essential factors for HCCI mixture formation were listed up as follows:

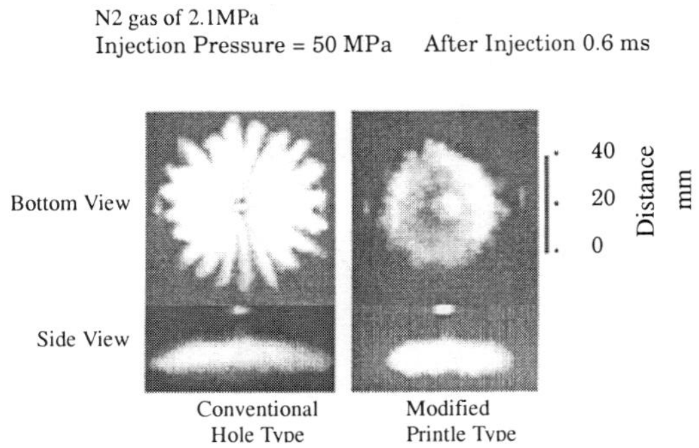

Fig.2 Spray Image of Experimental Injection Nozzle

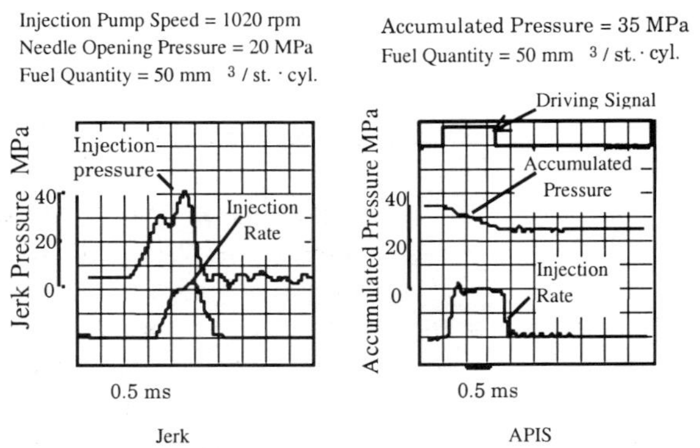

Fig.3 Injection Rate and Pressure

1. short injection duration, which was effective to make even distribution of fuel in space.
2. relatively low pressure injection, which contributed to suppress fuel vaporization in the nozzle.

The main object of mixture formation for HCCI engine is to build the homogeneous mixing condition of fuel from aspect of not only physical characteristics but also chemical characteristics. Once fuel is injected in the cylinder, chemical composition of fuel would be changed quickly during compression stroke of engine. If fuel is not enough mixed with air, there is the possibility to cause thermal cracking of fuel, which leads to extremely higher pollutant emissions. In order to know the process from onset of chemical reaction to ignition and combustion, combustion characteristics were examined on the diesel test engine of 2 liter displacement with this injection system.

COMBUSTION IMAGE OF UNIBUS

The combustion process was photographed with a high-speed camera, equipped with image intensifier and shadow graph system. Figure 4 shows the result.

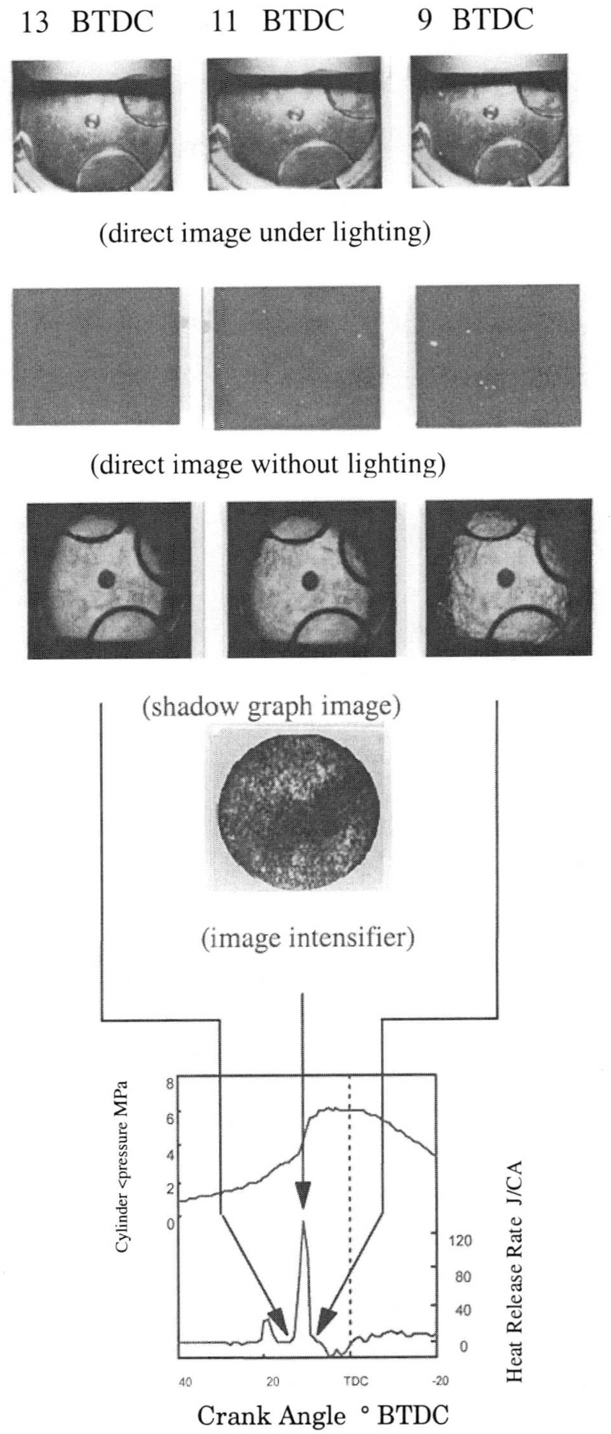

Fig.4 Combustion Image of UNIBUS

With ignition timing at 60BTDC, fuel was sufficiently dispersed, the spray core disappearing before ignition. At around 13BTDC, a dim shadow appeared, presumably corresponding to combustion. After the so-called "dormant" period, a clear shadow appeared at around 9BTDC. However, no luminous flame was observed in either shadow. Non-luminous flame portion was observed through the high-speed shutter TV camera with an image intensifier. The image for first heat release peak had very low photosensitivity. The image for the second peak, which appeared after the "dormant" period, had higher photosensitivity, and was photographed with the shutter speed decreased to about 1/300. As a result, although no luminosity was observed during its early stage of heat release, luminous spots were observed here and there after the heat release peak. This implies that the second heat release is due to mixture clouds scattered and combusted here and there at luminous spots. At either heat release peak, however, the luminosity is too weak to emit visible radiation.

EXHAUSTED GAS EMISSIONS

Figure 5 shows the emission characteristics and indicated mean effective pressure (IMEP) for various injection timings.

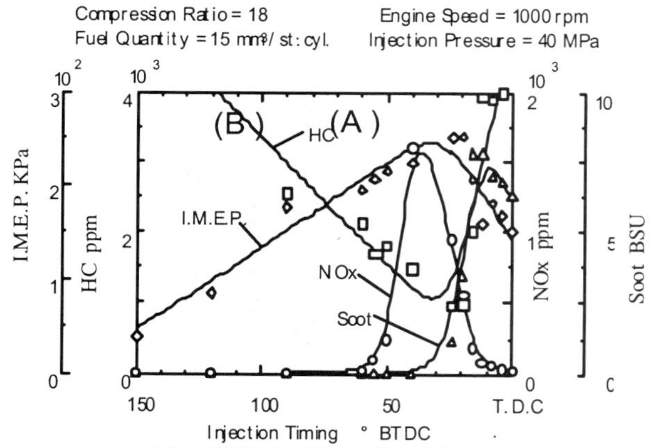

Fig.5 Effect of Injection Timing

As injection timing was advanced from the TDC, soot and HC decreased while NOx increased. IMEP also increased with the advancement of injection timing. As injection timing was advanced further, however, NOx and IMEP began to decrease while HC began to increase. There was an injection timing range in which NOx and soot emissions were nearly zero(region A). With injection timing in this range, auditorily evaluated combustion noise was extremely low, and IMEP decrease were relatively small. However, when injection timing was earlier than this range, IMEP decreasing became significant(region B). Using the injection timing at which both NOx and soot emissions became almost zero, the influence of fuel quantity on NOx emission was also determined. Figure 6 shows the result.

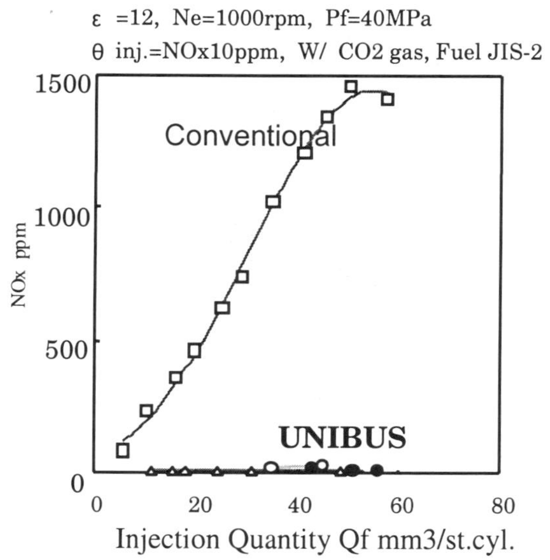

Fig.6 NOx emission according to loads

As fuel quantity was increased, NOx emission remained very low, about 1/100 that of the conventional DI diesel engine or lean-burn gasoline engine, if the oxidizer composition was diluted by mixing CO2 gas with air at the specific mixing ratio. When fuel quantity exceeded a certain level, however, NOx emission increased rapidly, nearing the level of the conventional DI diesel engine. Simultaneously, auditorily evaluated combustion noise became extremely high. With increasing EGR as a dilution gas, NOx emission increased at first but decreased at about 60% EGR. Fuel economy was improved according EGR rate, as shown in Figure 7. These facts imply that the combustion process changed suddenly when the oxidizer.

HEAT RELEASE RATE

The conventional diesel combustion system provided one heat release rate peak, whereas the new combustion system, with advanced timing of highly dispersed fuel injection provided two heat release peaks . This was understood that ignition depended on the extent to which injection timing was advanced, that is, the degree of fuel dispersion and the surrounding gas temperature. The first peak was thought as low temperature oxidation and the second as high temperature oxidation. EGR probably had a certain effect on ignition timing in the meaning of oxygen control. As the injection timing was advanced, generally heat release rate was split from the pattern with one peak into two parts.

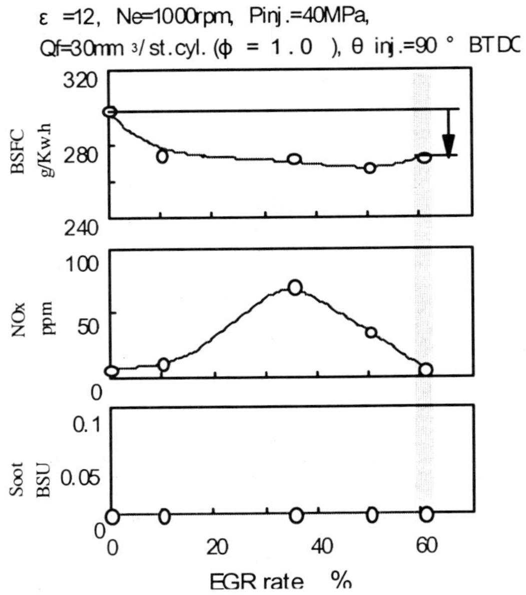

Fig.7 EGR Effect on Emissions and BSFC composition changed to a certain level

However, with heavy EGR rate of 60%, as shown in Figure 8, the two peaks changed to one peak and the peak generation was delayed to TDC direction. As the fuel cetane number became smaller, heat release peak generation was delayed and the rate of heat release became lower, but keeping two peak of heat release pattern, as shown in Figure 9. Fuel cetane number would not control the onset of ignition but chemical reaction speed. These interesting test results suggested that the equilibrium condition from slow to fast chemical reaction had an important meaning at ignition process.

TRIGGER OF IGNITION TIMING

In order to control ignition in the production engine with the conventional common rail injection, an idea of trigger for ignition was introduced. If some parts of mixture which plays as a trigger of ignition is completely managed in ignition, there would be a chance to build up time period where the other parts of mixture becomes homogeneous mixture.

Figure 10 shows the results under single and plural injection. Fig 11 shows exhaust gas emissions. At plural injection ,one was used as a trigger which led to ignition, and the other was as a main homogeneous mixture. The injection timing of second injection was set at a constant of 13ATDC. The chemical reaction rate was experimentally optimized by the first fuel injection. The test results showed that the reaction rate of low temperature oxidation varied according to the variation of injection timing. When the limited amount of fuel was injected in early stage of compression stroke, some weak chemical reaction was detected on heat release rate, as shown in Figure 12.

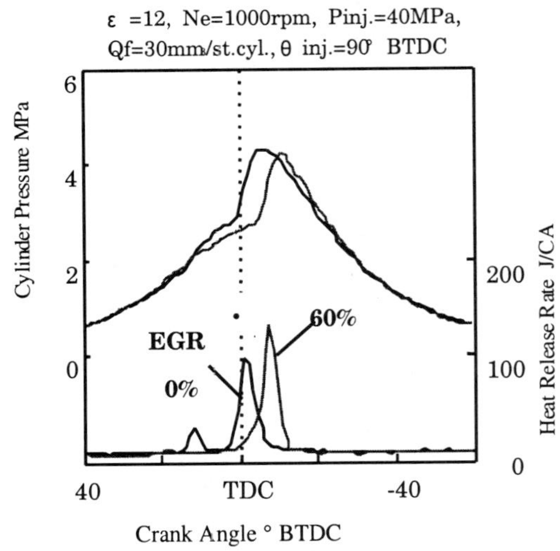

Fig.8 EGR Effect on Heat Release Rate

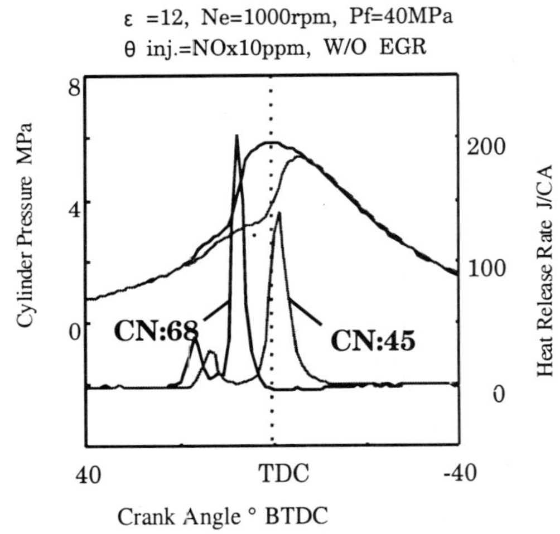

Fig.9 Fuel Cetane No. Effect on Heat Release Rate

However, it did not proceed to complete ignition, and torque of engine was not sufficiently obtained. When the second fuel injection was conducted after TDC of piston crank for keeping chemical reaction, the sufficient ignition and combustion was obtained with low pollutant emissions which led to enough torque production equivalent to conventional diesel.

OPTIMIZATION OF TRIGGER

Figure 12 showed the heat release rate with single/plural injection at various injection timing. The injection timing of main fuel was so retarded that combustion became unstable without trigger fuel. Furthermore, according to the timing of trigger injection, the heat release rate changed drastically. When the trigger fuel ignited before main fuel injection, the whole combustion deteriorated and emitted smoke and NOx.

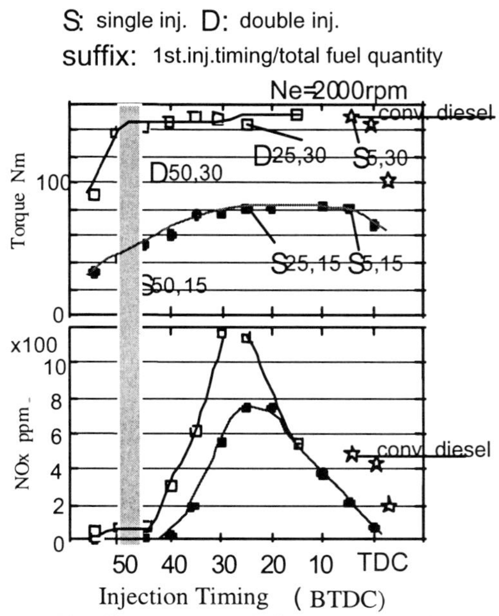

Fig. 10 Torque and NOx characteristics at single and plural injection

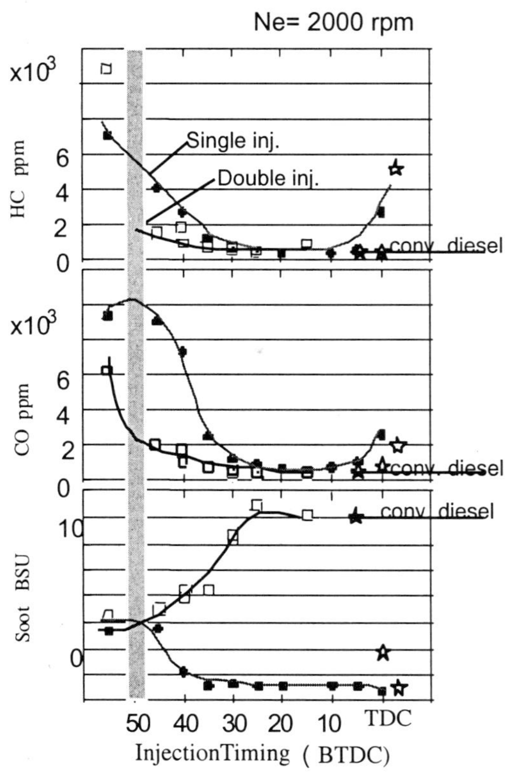

Fig. 11 HC,CO and Smoke characteristics

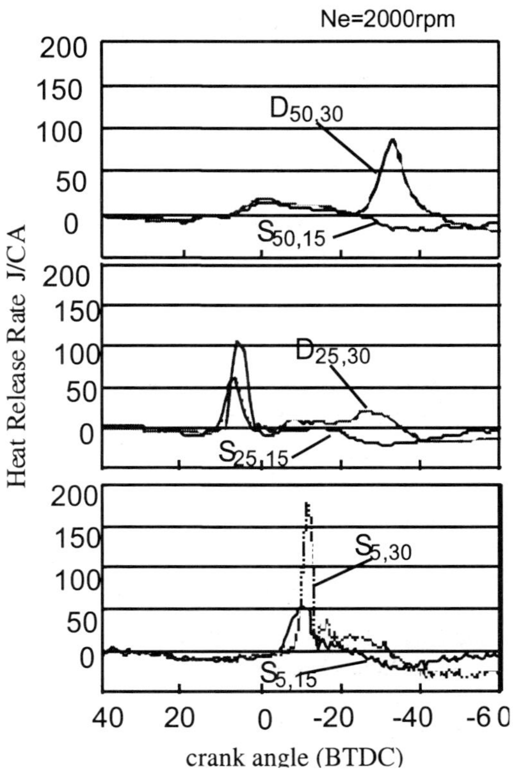

Fig. 12 Heat Release Rate

Figure 13 shows the mean gas temperature calculated from the measured cylinder pressure. When the pollutant emissions were controlled in very lower level (in case D50,30), the gas temperature was also suppressed in temperature close to 1500K. On the other hand, when the ignition control was not optimized for building long ignition delay, the gas temperature exceeded over 2000K where NOx was produced easily.

HOW TO CONTROL CHEMICAL REACTION

The major difference in combustion between the conventional diesel engine and HCCI engine seems to lie in the composition of fuel vapor at ignition[12]-[13]. In case of conventional combustion, the heat generated from the some local mixture parts caused thermal cracking of fuel in other fuel vapor and the detonative chemical reaction occurred. In case of combustion in HCCI, on the other hand, fuel oxidation occurred in the whole combustion space. Namely, chemical reaction started during the slow fuel oxidation process where fuel composition varied to the chemical species different from that in the conventional diesel combustion process. Usually, in HCCI, chemical reaction starts with low temperature oxidation of aliphatic fuel RH . The low temperature oxidation of the aliphatic fuel by the reaction O2 to produce the alkyl radicals R and HO2 is reported by

Prof. Dr. Warnatz in Heidelberg University (see Appendix). The alkyl radicals react with O2 to produce a alkylperoxy radical RO2 . If the temperature increases, RO2 radical will decompose back to the reactants. This leads to an inverse temperature dependence of the reaction. The second O2 addition path is the only one leading to chain branching. A distinct acceleration of the ignition process is expected by Warnatz as the results of the second O2 addition. Moreover, when thermo-dynamics equilibrium is considered in analysis of the above results, the following equations should be essential to control ignition.

$$R + O2 = RO2 \quad (1)$$

$$dlnK/dT = \Delta H/RT2 \quad (2)$$

$$dlnK/dP = -\Delta V/RT \quad (3)$$

here, ΔH means chemical reaction enthalpy and ΔV means volume of gas. These equations show the followings: If temperature increases, the equilibrium constant K decreases and the chemical reaction goes back to reverse direction. If pressure increases, the equilibrium constant K increases and equilibrium reaction shifts to higher temperature, that means, the low temperature oxidation remains even at higher temperature.

Namely, temperature performs as an acceleration work and pressure as a brake work at ignition process.

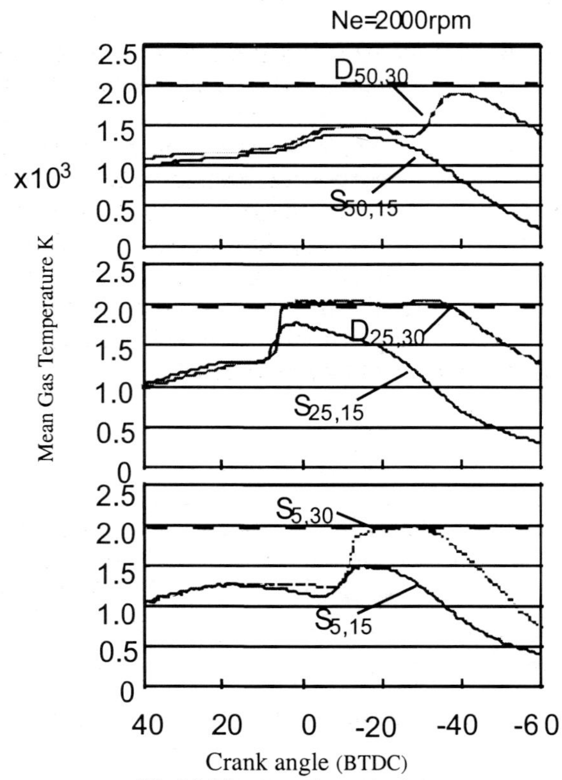

Fig.13 Mean gas temperature

TEMPERATURE AND PRESSURE EFFECT ON IGNITION

As above described, the fuel oxidation undergoes low temperature oxidation (LTO) first, and then shifts to high temperature oxidation (HTO). Figure 14 shows the ignition process with time schematically. The time interval involved in the transition from LTO to HTO is determined by the heat release quantity of LTO. When the temperature rises above the specific level due to LTO, or the surrounding temperature is sufficiently high beyond the limit temperature, LTO stops promptly and HTO accelerates.

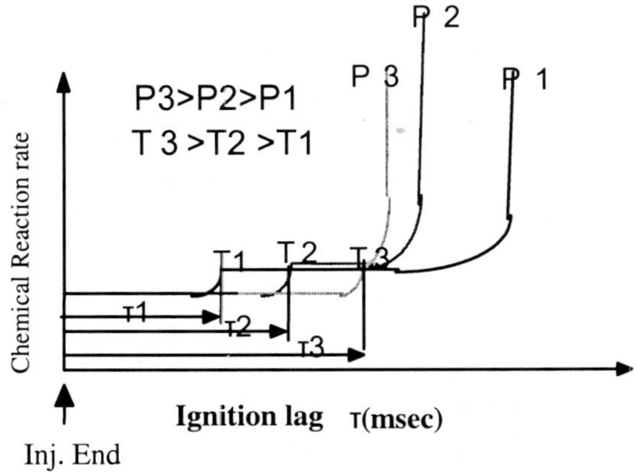

Fig.14 Ignition process with time

However, when the surrounding pressure is higher, LOT remains even at higher temperature level and promptly shifts to HTO as illustrated in Figure.14. Conclusively, the pressure is higher, the ignition lag becomes longer. During the extended period of ignition lag τ, mixing of fuel with air proceeds to great extent. In Figure 15 ,the combustion images are shown. One is the combustion with the controlled ignition, the other the combustion without controlled ignition, respectively. In case of the well-controlled combustion , homogeneous mixture was formed by the combination of plural injections, and combustion image showed an invisible flame.

Thus, the controlling for ignition lag have a decisive meaning for HCCI engine. Figure 16 shows ignition lag characteristics according to temperature and pressure schematically. As illustrated in the figure, negative temperature coefficient(NTC) region shifts to higher temperature with increasing pressure. If pressure decreases after TDC crank angle, NTC point shifts to lower temperature again. It means that there is a possibility to switch from LTO to NTC region leading to HTO by reducing pressure in the expansion stroke of engine .

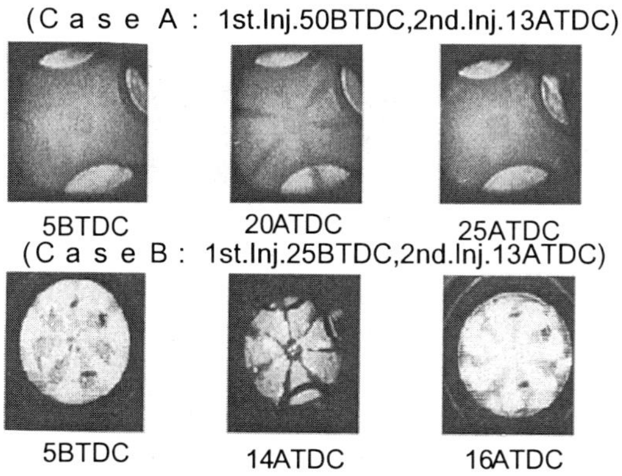

Fig.15 Combustion image underwith/without control

APPLICATION

The new concept was introduced in the vehicle (I.W.2250Kg) with new DI diesel engine (1KD-FTV,3liter-4cylinder) equipped with common rail injection system and conventional hole type nozzle, in August 2000 in Japanese market . Low pollutant emissions and low noise characteristics was verified by the engine test. The Map of operating with UNIBUS where NOx and Smoke are reduced extremely are shown in Figure17.

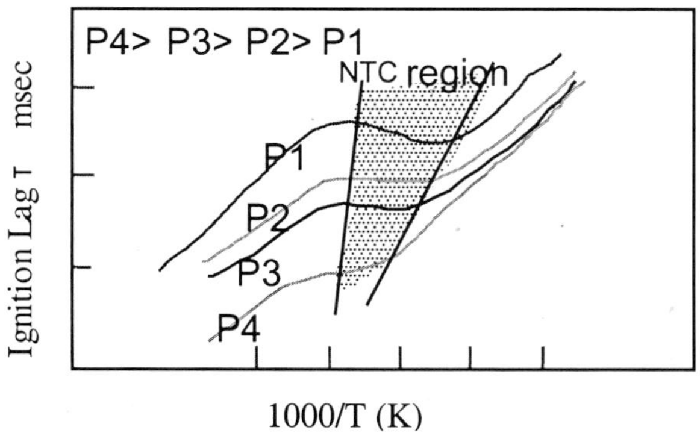

Fig.16 Ignition delay with pressure dependence

In UNIBUS operation, very low NOx emission below 70ppm and soot close to zero was realized in Japanese test mode. In higher loads beyond UNIBUS operation, early pilot operation was conducted where NOx level is similar to that of conventional diesel and Smoke is considerably reduced. At lower speed operating conditions, smoke did not produced even at equivalence ratio of about 1. At higher speed operating condition, the conventional type hole nozzle was not enough to perform making homogeneous mixture. Presently, UNIBUS operation is restricted within the half loads when UNIBUS concept is introduced simply without modification of injection system. The advantage of fuel economy is now about 13% compared with IDI diesel .

DISCUSSIONS

The subjects of HCCI is now how to expand UNIBUS operation range, especially how to make homogeneous mixture in higher engine speed operating conditions. A clue to the subject is micro-hole nozzle. The down sizing of nozzle hole size combined with increasing number of holes showed a good tendency for improving mixture in the engine test.

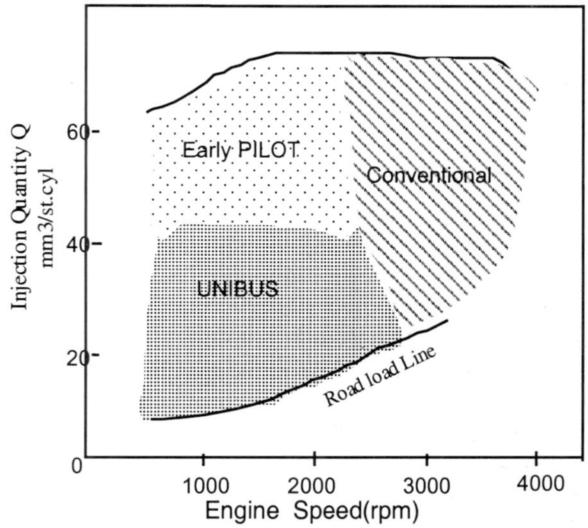

Fig.17 Operation Map with UNIBUS (HILUX SURF, LAND CRUISER PRADO 2250Kg WT)

Other possibility is to control mixture temperature which relates closely with evaporation of fuel. Therefore, evaporation mechanism of multi-composition fuel should be investigated further. Also the low gas temperature combustion of HCCI engine would lead after treatment systems to a new generation. For instance, Soluble Organic Fraction produced easily by HCCI combustion is expected to react with NOx. Further possibility of HCCI is to operate engine under fuel-air equivalence ratio of 1, without exhausting black smoke. 18.

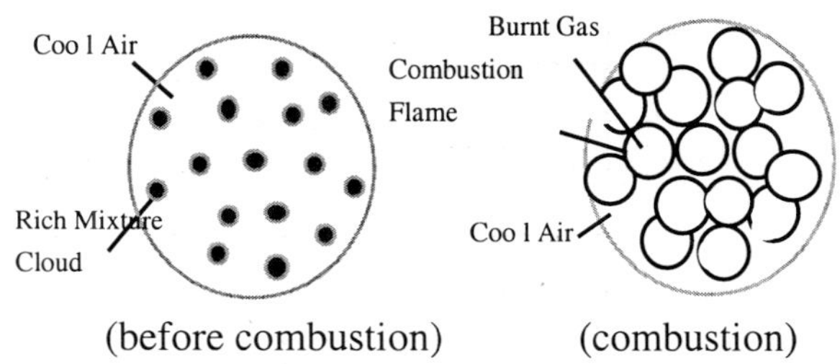

Fig. 18 Homogeneous-stratification for HCCI

This means that the after treatment system for gasoline engine can be applied for diesel engine. Moreover, the definition of "homogeneous mixture" should be made clear when HCCI will grow up to be a new generation of engine combustion process. Author will recommend it as homogeneous-stratification, shown in Figure 18. Mixture ratio in diesel engine should be kept within the limits to avoid misfiring . It is supposed as a goal that chemical reaction proceeds simultaneously in each minimum unit of fuel cloud distributed over whole combustion space, keeping their temperature to be equilibrium one.

CONCLUSIONS

With understanding of chemical reaction occurred in HCCI engine in details, the combustion of HCCI came up with real images. The ignition control should be made not only from physical aspect, but also from the combined aspects of chemical and physical phenomenon.

1. Ignition timing control for HCCI engine was realized by preparing ignition trigger injection.
2. The trigger injection enabled to keep low temperature oxidation which extended the ignition lag.
3. During the extended ignition lag, whole quantity injected dispersed in the combustion space and became homogeneous mixture.
4. Equilibrium condition in low temperature oxidation has the decisive meaning in HCCI combustion process.

REFERENCES

1. Soemya(Ed), T., Advanced Combustion Science, Springer Verlag,1993
2. Chevalier,C., et al, Hydrocarbon Ignition Automatic generation of reaction Mechanisms and applications to modeling of Engine knock, Twenty Fourth Symposium(International) on Combustion / The Combustion Institute, 1992/pp.93-101
3. Miwa.K. et al, JSME. Int., Vol31.No.1 (1988), 166
4. Uyehara,O,A., Reducing NOx in ICE,SAE Paper 940896 (1994)
5. Yanagihara, H. Sato, Y. Y.,Sakakibara, Y.,Mizuta, J.,Kawamura,K and Kataoka,M., An Analysis of Fuel Spray and Combustion at DI Diesel Engine with Accumulated Press. Injector, JSAE Symposium, No.912(1991-10)
6. Takeda,Y, et al., Emission characteristics of lean premixed diesel combustion with an advanced fuel injection timing, The 73rd JSME Fall Annual Meeting,Vol.3,pp.188- 191(1995)
7. Aoyama,T., et al., An experimental study on premixed charge compression ignition gasoline engine, SAE Paper 960081(1996)
8. Yanagihara,H.,A simultaneous reduction of NOx and soot in diesel engines under a new mixture preparation, The 73rd JSME Fall Annual Meeting,Vol.4,F5-(1)(1995)
9. Yanagihara,H. et al, A simultaneous reduction of NOx and soot in diesel engines under a new combustion system (uniform bulky combustion system UNIBUS), 17th. Int. Vienna Motor Symposium (1996)
10. Sato,Y. et al., A simultaneous reduction of NOx and soot in diesel engines under a new combustion system, Proceedings of JSAE, 9633946, pp.81-84(1996)
11. Yanagihara,H,A, Study of DI Diesel Engine Combustion under Uniform and Highly Dispersed Mixture Formation-Possibility of Ignition timing Control-,Proceedings of JSAE,9830956
12. Maria, N.et al, Proc.Combustion.Inst.26.773-780 (1997)
13. Warnatz, J.et al, Combustion, Second Edition, Springer-Verlag (1999)

APPENDIX

LOW TEMPERATURE OXIDATION MECHANISM-[13] The sequences of chemical reaction in low temperature oxidation proceeds as follows:

RH aliphatic fuel

R· alkyl radicals

$R\cdot + O_2 \rightleftarrows RO_2\cdot$ (1st. O2 addition)

$RO_2\cdot \rightarrow QOOH\cdot$

(Internal H atom abstraction)

$QOOH\cdot + O_2 \rightleftarrows O_2QO_2H\cdot$

(2nd. O2 addition)

$O_2QO_2H\cdot \rightarrow HO_2Q'O_2H\cdot$

(Internal H atom abstraction)

$HO_2Q'O_2H\cdot \rightarrow HO_2Q'O + OH\cdot$

(chain propagation)

$HO_2Q'O \rightarrow OQ'O\cdot + OH\cdot$

(chain branching)

$OQ'O\cdot \rightarrow RCHO$

(decomposition)

$RCHO + X \rightarrow RCO\cdot + XH$

$RCO\cdot + O_2 \rightarrow RCOO_2H\cdot$

$RCOO_2H\cdot \rightarrow R\cdot + CO_2 + OH\cdot$

(decomposition)

$RCO\cdot \rightarrow R\cdot + CO$ (decomposition)

A New Generation of Engine Combustion Processes for the Future?

27 rue Ginoux, 75015 Paris

NEW DIESEL COMBUSTION PROCESS TO ACHIEVE NEAR ZERO NOx AND PARTICULATES EMISSIONS

Bertrand Gatellier, Bruno Walter, Marjorie Miche
IFP, France

ABSTRACT

The paper describes the design considerations and test results of the NADI (Narrow Angle Direct Injection) concept recently developed at IFP. This "dual mode" engine application applies new combustion processes such as Homogeneous Charge Compression Ignition (HCCI) at part load and switches to conventional Diesel combustion to reach full load requirements. The results indicate that the NADI system could be an answer facing the Diesel engine challenges (bring local pollution to the level requested by city air quality standard and reduce CO_2 emissions in order to minimize the global warming risk). The use of HCCI combustion mode allows near zero particulate and NOx emissions and maintains very good fuel efficiency. By using conventional Diesel combustion mode, NADI system is consistent with future Diesel engine power density standard. Further development steps are also discussed at the end of the paper.

INTRODUCTION

Thanks to their high thermal efficiency coupled with low CO_2 emissions, Diesel engines are promised to an increased part of the transport market if their NOx and Particulate emissions are reduced. Today, adequate after-treatments NOx and PM traps are under industrialization with still concerns about fuel economy, robustness, sensitivity to fuel sulfur and cost.

New combustion process such as Homogeneous Charge Compression Ignition (HCCI) are investigated for their potential to achieve near zero particulate and NOx emissions [1,12]. They constitute alternative or complementary answers to the sophisticated and complex after-treatments strategy. Their principle consists in preparing a highly diluted by burned gases air/fuel mixture, in achieving its simultaneous ignition in the whole space of the combustion chamber and in precisely controlling such combustion for the best performance in terms of efficiency and pollutant emissions. Because the combustion takes place in a homogeneous way throughout the bulk of the mixture, the thermal NOx formation and the soot production are in principle much less than with the diffusion flame typical of the conventional Diesel. However, there are some difficulties in applying HCCI combustion to engines. The main drawbacks are too high hydrocarbons (HC) and carbon monoxide (CO) emissions, poor combustion control at high load and then limited operating range and power output.

The ambitious goal of this work is to develop a combustion system able to reach near zero particulate and NOx emissions while maintaining the higher and higher performance standards of the D.I. Diesel engines.

IFP APPROACH DESCRIPTION

IFP CONCEPT OVERVIEW

To overcome the limitations in power output, IFP has developed a "dual mode" engine that uses HCCI combustion at low and medium loads and conventional Diesel combustion at high loads. That means that the combustion system should be able to switch between the two combustion modes.

At an early development stage, we have decided to keep the general architecture of a conventional Diesel combustion system:

- Direct injection
- Flat cylinder head
- Combustion piston bowl

A common rail fuel injection system has been selected due to its continuously increased flexibility, especially in terms of injection events.

A narrow spray cone angle was selected (less than 100 degrees) to limit fuel wall wetting, while having a great flexibility in terms of injection timing (very early injection).

Concerning the valve control system, variable valve actuation is used to adjust the effective compression ratio for better combustion control. However, 1D simulations have shown that mechanical hardware could be sufficient.

Concerning the compression ratio, a lot of work pointed out the advantage to have a moderate compression ratio in order to better control the start of combustion, especially to extend the engine HCCI operating range. In the present study, the engine geometric compression ratio has been set to 16:1. This choice seems to be consistent with engine cold starting because the NADI concept use a VVA system and a narrow injection angle. The early closure of the intake valve(s), around BDC, during engine cranking, allows an increase of effective compression ratio by one unit. At the same time, it will be possible to optimize the injections events during engine cranking, without wall wetting drawbacks thanks to the selected narrow injection angle.

Figure 1 give an overview of the NADI combustion system concept whose main features can be listed as:

- Conventional flat cylinder head
- Narrow spray cone angle
- Reduced geometric compression ratio (16:1)
- Multi stage injection (Common rail FIS)
- VVA hardware

COMBUSTION SYSTEM DEFINITION

Our first objective was to find a combustion chamber well adapted to a narrow spray cone angle in order to reach a good conventional Diesel combustion behavior, especially at full load. The IFP approach is based on the use of CFD tools combined with development work done with an

FIG.1

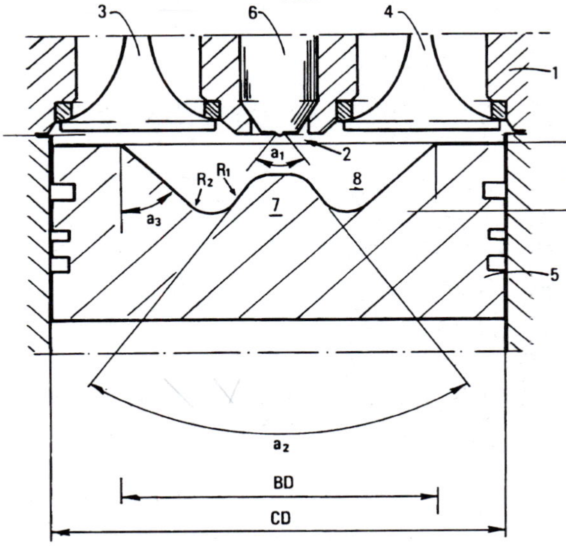

Figure 1: Overview of the IFP combustion system concept

operating engine. A brief description of the CFD tools is given in appendix.

The first challenge was to make a proper transport and mixing of with air the fuel injected inside the piston bowl. Several combustion chamber shapes associated with nozzle geometry and swirl motion variations have been computed. The computed output power against burned mass fraction just before the exhaust valve opening is shown in figure 2. After few iterations, some combustion chambers, associated with nozzle and swirl definition, adapted to narrow spray cone angle have been defined.

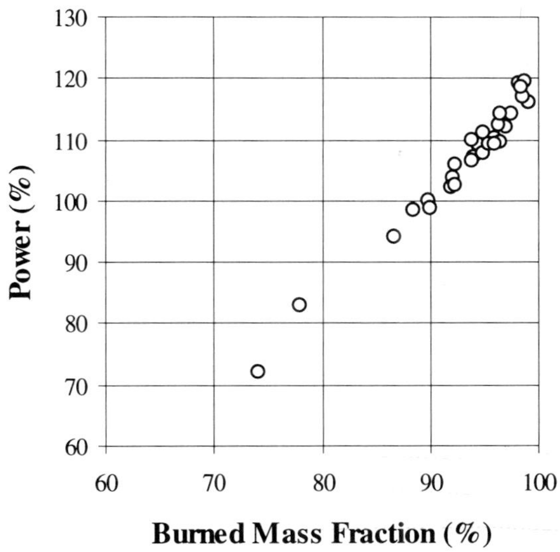

Figure 2: CFD results at 4000 RPM, full load

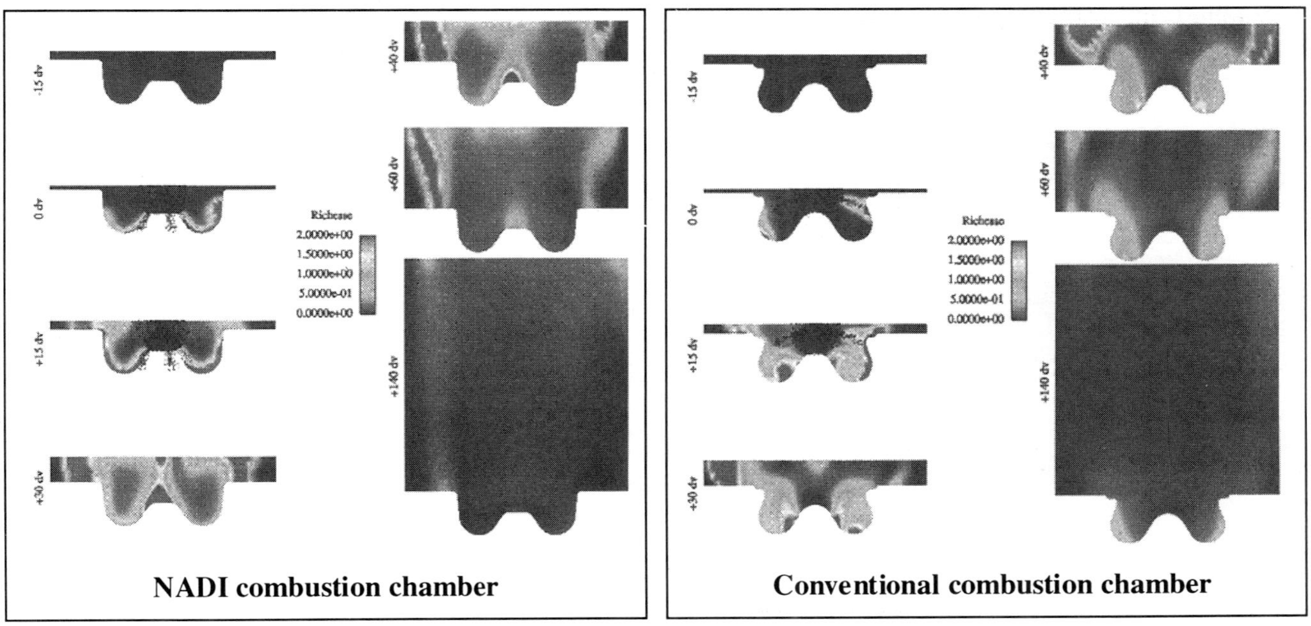

Figure 3: Computed combustion process at 4000 RPM, full load

CFD computations helped to define the best fuel injection and combustion process. Figure 3 shows the combustion process at 4000 RPM, full load, with a conventional nozzle and combustion chamber geometry and with the NADI concept. With conventional combustion system, the fuel is injected towards the bowl periphery. Due to fuel/wall interaction, the majority of the fuel is transported to the center of the bowl, mixes with air and burns. Some fuel mixes with air and burns in the squish area. With the NADI concept, the fuel is injected at the center of the combustion piston bowl. Due to fuel/wall interaction, fuel is transported to the piston bowl periphery, mixes with air and burns.

FIRST ENGINE EVALUATION

The first target was to validate the NADI concept on a single cylinder engine with conventional combustion mode at full load.

ENGINE DESCRIPTION

The experiments were performed on a single-cylinder engine. The main characteristics are given in table 1. The intake ducts have been modified in order to adapt the swirl motion. All the results shown in this paper, including full load conditions have been obtained with the same swirl number. A production Bosch common rail fuel injection system (maximum injection pressure: 135 MPa) has been modified in order to allow up to eight injections per cycle. The engine was externally boosted and the exhaust pressure was controlled by a throttle valve in accordance with intake pressure and EGR values.

Bore	78.3 mm
Stroke	86.4 mm
Displacement	416 cm^3
Compression ratio	16:1

Table 1: Engine data

INITIAL RESULTS AT FULL LOAD

The first target for the NADI concept development program was to find the same performance, in terms of power, as for the initial conventional combustion system. To have a fair comparison basis, the original combustion piston bowl was modified in order to reduce the compression ratio to 16:1. With the following limiting factors and parameters settings, a power of 45 kW/liter[1] was achieved at 4000 RPM, full load.

- Maximum in-cylinder pressure: 15 MPa
- Exhaust temperature: 750°C (equivalent to 710°C on a single cylinder)
- Smoke: < 3 FSN
- Intake pressure: 2050 hPa
- Exhaust pressure: 3150 hPa
- Intake temperature: 60°C

The conventional combustion system and the first version of the NADI concept were coMPared on the same single-cylinder engine at 4000 RPM, full load. The figure 4 shows the exhaust temperature and the smoke evolution during a fuel injected quantity increase. NADI combustion system shows good conventional Diesel combustion behavior, reaching almost the same power (only 1% less) with more smoke, but still below the limit. This first engine evaluation had confirmed the possibility to find a combustion chamber adapted to narrow spray cone angle. Nevertheless, some drawbacks were pointed out at this initial development stage. We have observed an increase of the ignition delay and a smaller reaction rate. So, these preliminary results have been obtained with an advanced injection timing, with some disadvantages on maximum in-cylinder pressure (+0.5 MPa), on fuel consumption (+10%) and on combustion noise (+2dB). As it will be discussed later, improvements have been done to avoid these drawbacks.

[1] Calculated from IMEP, with a friction mean equivalent pressure (FMEP) of 2 bar (FMEP).

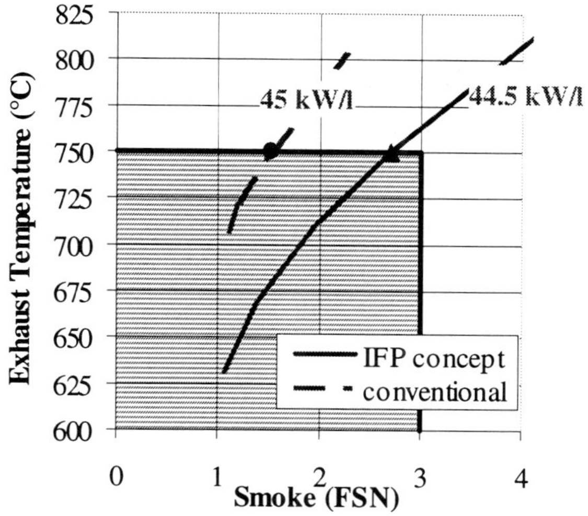

Figure 4: First results at 4000 RPM, full load

COMBUSTION SYSTEM IMPROVEMENT

After the first evaluation of the NADI concept, the combustion system has been improved to obtain better results at high load, using conventional combustion mode. After that, the NADI concept has been evaluated at part load using HCCI combustion mode.

IMPROVEMENTS AT FULL LOAD

The future Diesel engine requirements in terms of power density (up to 60 kW/liter) will lead to higher intake pressure and in-cylinder pressure. In this context, it becomes obvious that a too high sensitivity to injection timing could be a critical point. The combustion bowl shape has been modified in order to reduce the ignition delay and to improve the combustion rate by promoting a better air/fuel mixing.

With the improved design, the combustion behavior is less sensitive to injection timing variations and it is possible to reach the future Diesel engine requirements in term of power density. Figure 5 shows the improvements realized. The intake pressure has been increased up to 3000 hPa at a constant maximum in-cylinder pressure of 16 MPa.

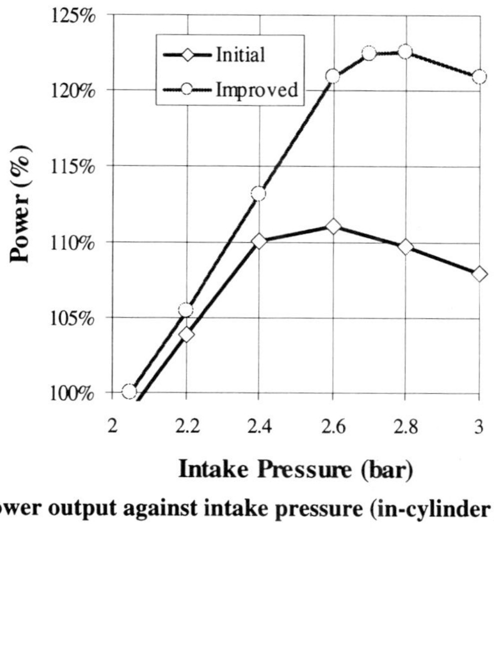

Figure 5: Engine power output against intake pressure (in-cylinder pressure < 16 MPa)

RESULTS IN HCCI MODE

Two engine speeds have been selected to perform our first HCCI mode engine evaluation (1500 RPM and 2500 RPM). At this initial stage, two geometric compression ratios (16:1 and 14:1) have been used to easily evaluate its effect on combustion ignition. In the meantime, 1D calculations have shown that it is possible to adjust pressure and temperature evolution against crank angle by varying the intake valve closure at constant geometric compression ratio of 16:1. At part load, the NADI concept (using HCCI mode) is coMPared with standard geometry (using conventional combustion, a compression ratio of 18:1 and parameters setting consistent with Euro III emission standard).

Results at 1500 RPM

The figures 6 to 8 show fuel consumption and pollutant emissions against engine load. External EGR was used to control the ignition of the combustion. Only a light boost pressure has been used above 5 bar of IMEP with the lowest compression ratio.

Operating range and fuel consumption: With the geometric compression ratio of 16:1, it is possible to achieve 4 bar of IMEP using HCCI mode. The fuel consumption increase is due to the unburned fuel in the exhaust (HC and CO), which could represent up to 10% of the injected fuel at a IMEP of 4 bar. As figure 9 shows, air dilution by exhaust gas increases the ignition delay and allows a good combustion timing control. However, an increase of EGR decreases the air/fuel ratio and could, above a certain limit, increase the unburned fuel with a negative iMPact on fuel consumption. With the geometric compression ratio of 14:1, it is possible to reach 6 bar of IMEP using HCCI mode. The fuel consumption is better because less EGR is needed to properly control the combustion timing with a net benefit on the unburned fuel which represents only 2% of the injected fuel quantity at 5 bar of IMEP.

Emissions: Concerning NOx and particulate emissions, as the figure 7 shows, the levels are near zero (always below 0.03 g/kWh) and consistent with SULEV standard levels. NOx emissions are more than 100 times lower than conventional Diesel engine and particulate emissions are more than 10 times lower. The HC and CO emissions are not too high with levels in the same range or lower than direct gasoline engine one's. However, as the exhaust temperature is low, especially at low load, improved oxidation catalyst will be required.

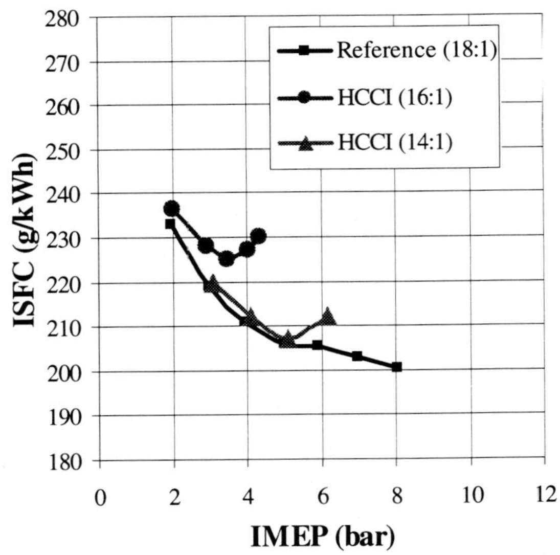

Figure 6: ISFC against IMEP at 1500 RPM

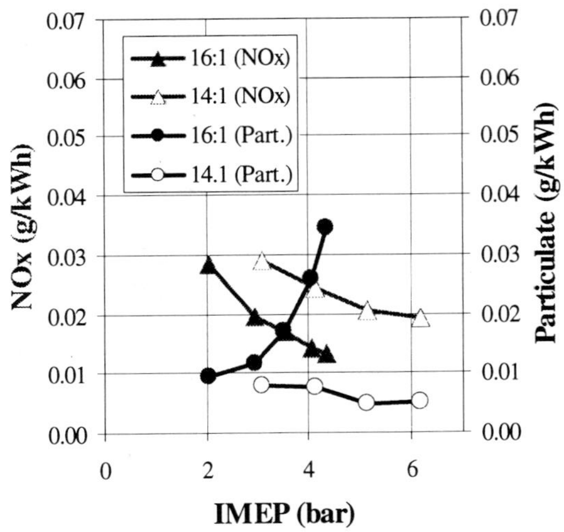

Figure 7: Emissions against engine load at 1500 RPM

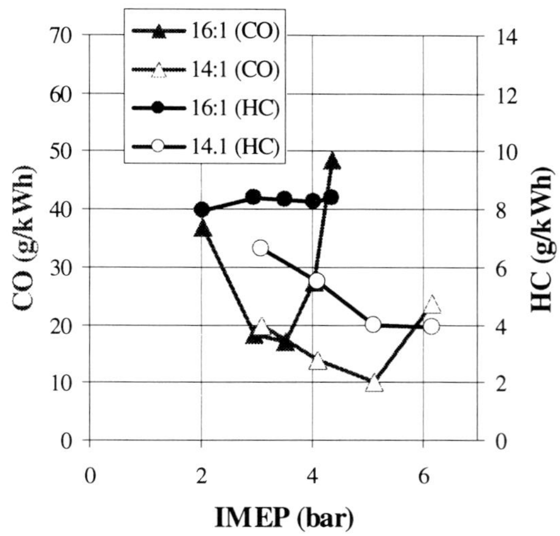

Figure 8: Emissions against engine load at 1500 RPM

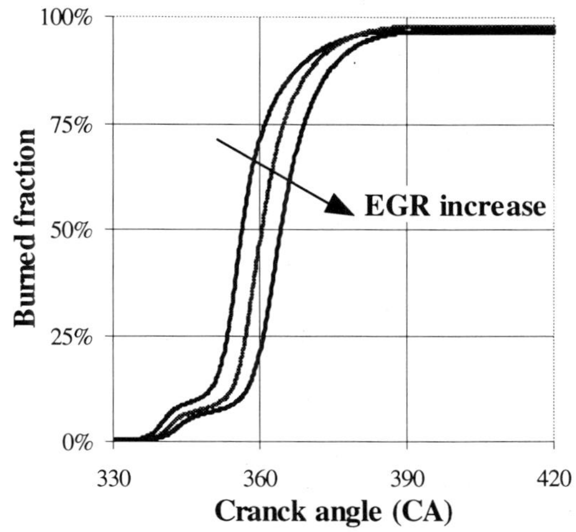

Figure 9: Influence of EGR on combustion timing

Results at 2500 RPM

The figures 10 to 12 show fuel consumption and pollutant emissions against engine load. External EGR was used to control the ignition of the combustion. At high load, the EGR flow has been cooled in order to maintain the recycled gas temperature below 90°C. The intake pressure has been progressively increased for IMEP higher than 5 bar. Notice that the boost pressure has been always less than conventional value in order to take into account the less energy available on the turbine, due to HCCI combustion mode.

Operating range and fuel consumption: Due to higher engine speed, it is possible to achieve good HCCI combustion mode at higher load. It is possible to achieve 6 bar of IMEP with a compression ratio of 16:1, and 9 bar of IMEP with a compression ratio of 14:1. using HCCI mode.

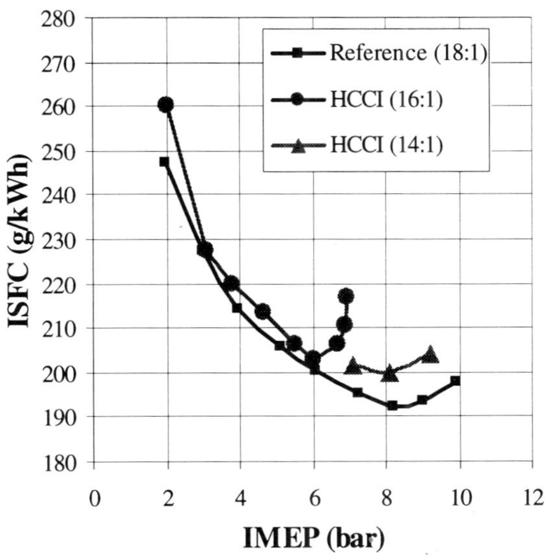

Figure 10: ISFC against IMEP at 2500 RPM

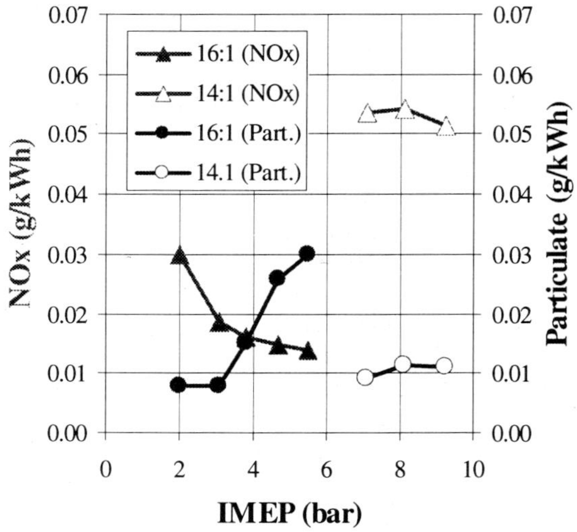

Figure 11: Emissions against engine load at 2500 RPM

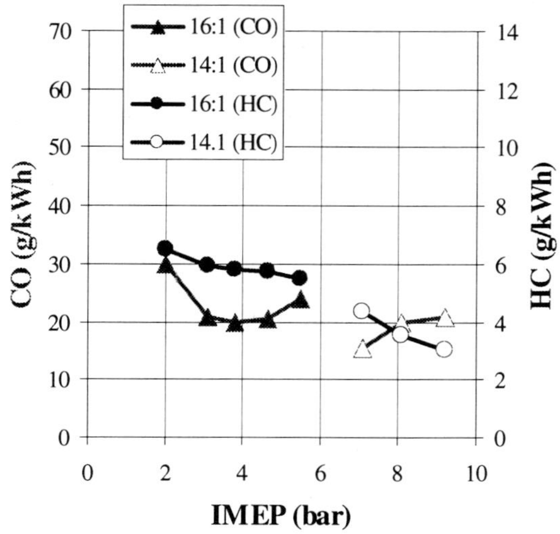

Figure 12: Emissions against engine load at 2500 RPM

Emissions: Once again, near zero NOx and particulate emissions are achieved as shown in figure 11. The HC and CO emissions are in the same range of direct gasoline engine. At this engine speed, the exhaust temperature is high enough to more easily oxidize the HC and CO by a catalyst. Depending on engine parameters settings such as EGR or injection timings, it is possible to find different trade-off between the four pollutants. The results shown with the lowest compression ratio illustrate this point.

PERSPECTIVES

The NADI concept has already achieved interesting results. 50 kW/l has been reached with conventional limiting factors and engine parameters setting (in-cylinder pressure below 15 MPa, exhaust temperature below 750°C, smoke less than 3, intake pressure of 2300 hPa). In order to fulfill the future D.I Diesel engine standard, directions of progress are identified.

To improve the conventional combustion mode, the increase of injection pressure (from 135 MPa to 160 MPa or higher) will promote the air/fuel mixing. The nozzle configuration (number and diameter of the holes) could be optimized by taking into account the completely different trade-off between pollutant emissions at low and medium load and power at full load.

To improve the HCCI combustion mode, the control parameters settings such as the split of injection, the exhaust gas recirculation or the boost pressure has to be deeply optimized. The iMPact of new technologies such piezo-electric common rail FIS to increase the number of injection events, electric assisted turbo charger to enlarge the operating range or for advanced development fully VVA system, has to be precised.

Due to the complexity of the processes involved, IFP has already planned to use sophisticated scientific tools, such as optical access engines and associated diagnostics and 3D CFD tools, to get the knowledge necessary to support the NADI concept further development process to the industrial application.

CONCLUSION

This paper presented the main design features and preliminary results of the NADI new combustion system applied to an existing D.I Diesel engine. It was shown that this concept could be an answer to the Diesel engine challenges. This "dual mode" engine has reached near zero NOx and particulate emissions at part load (including MVEG operating range) by using some homogeneous charge compression ignition combustion and is coMPatible with further specifications concerning power density.

Some points have to be addressed and further development works are already planned.

- CO and HC emissions have to be improved, especially at low engine speed and engine load.
- Transition strategies between the two mode have to be defined, with a special attention to EGR control
- HCCI engine load domain has to be little extended

Nevertheless, advanced engine technologies such as VVA (Variable Valve Actuation), low temperature oxidation catalysts light off, electric assisted turbocharger or further generation of common rail fuel injection system make realistic practical applications of NADI engine in near future production powertrains

ACKNOWLEDGMENTS

The authors would like to express their acknowledgements to Pascal Gréau for diligently running the engine experiments.

REFERENCES

1. -R. H. Stanglmaier and C. E. Roberts « Homogeneous Charge Compression Ignition (HCCI): Benefits, Compromises, and Future Engine Applications » SAE Paper 1999-01-3682, (1999). Southwest Research Institute
2. -G. R. Pucher, D. P. Gardiner, M. F. Bardon, V. Battista « Alternative Combustion Systems for Piston Engines Involving Homogeneous Carge Compression Ignition Concepts – A review of Studies Using Methanol, Gasoline and Diesel Fuel » SAE Paper 962063 (1996).
3. -P. M. Najt, D. E. Foster « Compression – Ignited Homogeneous Charge Combustion » SAE Paper 830264, (1983). University of Wisconsin – Madison
4. -M. Christensen, A. Hultqvist and B. Johansson « Demonstrating the Multi Fuel Capabilityof a Homogeneous Charge Compression Ignition Engine with Variable Compression Ratio » SAE Paper 1999-01-3679, (1999). Lund Institute of Technology
5. M. Christensen and B. Johansson « Supoercharged Homogeneous Charge Compression Ignition (HCCI) with Exhaust Gas Recirculation and Pilot Fuel » SAE Paper 2000-01-1835, (2000). Lund Institute of Technology
6. -H. Yokota, H. Nakajima and T. Kakegawa « A new Concept for Low Emission Diesel Combustion (2nd Rep.: Reduction of HC and CO Emission, and Improvement of Fual Consumption by EGR and MTBE Blended Fuel » SAE Paper 981933, (1998). Hino Motors Ltd
7. -H. Yanagihara, Y. Sato, J. Mizuta « A simultaneous reduction of NOx and soot in diesel engines under A new combustion system (Uniform Bulky Combustion System – UNIBUS » VDI 1996 n° 267, (1996). Toyota Motor Corporation and Central R&D Labs
8. -Y. Iwabuchi, K. Kawai, T. Shoji and Y. Takeda « Trial of New Concept Diesel Combustion System –Premixed Compression – Ignited Combustion » SAE Paper 1999-01-0185, (1999). Mitsubishi Motors Corporation
9. -T. Hashizume, T. Miyamoto, H. Akagawa and K. Tsujimura, "Combustion and Emission Characteristics of Multiple Stage Diesel Combustion." SAE Paper 980505, (1998). New A. C.E. Institute Co. Ltd
10. -N. Shimazaki, H. Akagawa and K. Tsujimura « An Experimental Study of Premixed Lean Diesel Combustion » SAE Paper 1999-01-0181, (1999). New A. C.E. Institute Co. Ltd
11. S. Kimura, O. Aoki, Y. Kitahara and E. Aiyoshizawa « Ultra – Clean Combustion Technology Combining a Low – Temperature and Premixed Combustion Concept for Meeting Future Emission Standards » SAE Paper 2001-01-0200, (2001). Nissan Motor Co, Ltd
12. - M. Odaka, H. Suzuki, N. Koike, H. Ishii « Search for Optimizing Method of Homogeneous Charge Diesel Combustion » SAE Paper 1999-01-0184, (1999). Traffic Safety and Nuisance Research Institute

CONTACT

bertrand.gatellier@ifp.fr

APPENDIX

The KMB code, which stands for Kiva-Multi-Block, is based on the KIVA-II code originally from Los Alamos Laboratory [ref1]. It makes possible computation of fully 3 dimensional, transient, compressible, chemically reactive, turbulent flows in moving geometries, with sprays. The formalism of the equations, the averaged Navier-Stokes equations coupled with turbulence equations, is an Arbitrary Lagrangian Eulerian finite volume formalism, resolved on hexahedrons in structured domains.

A multi-block version developed at IFP [ref2], allows computation in complex geometries. Each part of a geometry, for example the cylinder and the intake port, can be meshed and logically structured independently. Some overlap allows physical data to be transmitted from one block to the other. This method drastically reduces the number of ghost cells with respect to a single block containing all the geometry.

A remapping algorithm is also used at valve closing, to transfer all the computed fields from a grid with valve used for intake phase, to a grid without a valve used for compression and combustion phases. This avoids very thin and distorted cells near the engine head during these two last phases. This more homogeneous grid allows larger time steps and better accuracy during injection and combustion. These abilities are described in [ref3].

The calculations presented here use the following models for describing the physical processes occurring within the combustion chamber: a standard κ–ε model for turbulence, a classical logarithmic law of the wall, two models for Diesel combustion (a kinetic combustion model for self-ignition and the Magnussen model [ref4] for turbulent combustion), and the Diwaker model [ref5] for heat transfer.

[ref1] Amsden A.A., O'Rourke P.J., Butler T.D., "*KIVA-II: a Computer Program for Chemically Reactive Flows with Sprays*", Los Alamos LA-11560-MS, 1989.

[ref2] Habchi C. Torres A., "*A 3D Multi-Block Structured Version of the KIVA 2 Code*", First European CFD Conference proceedings, pp 502-512, 1992.

[ref3] Torres A., Henriot S., "*3D Modeling of Combustion in Lean Burn Four Valves Engines: Influence of Intake Configuration*", 3th International Symposium on diagnostics and Modeling of combustion in Internal Combustion Engines, COMODIA 1994.

[ref4] Magnussen B.F., Hjertager B.H., "*On Mathematical Modeling of Turbulent Combustion with Special Emphasis on Soot formation and Combustion*", 16th Symposium on Combustion, Combustion Institute, pp719-729, 1976.

[ref5] Diwakar R., "*Assessment of the Ability of a Multidimensional Computer Code to Model Combustion in a Homogeneous Charge Engine*", SAE paper 840230.

GASOLINE
CAI & HCCI COMBUSTION

A New Generation of Engine Combustion Processes for the Future?
© P. Duret (Editor) and Editions Technip, Paris, 2001, pp. 55 - 68
27 rue Ginoux, 75015 Paris

THE HETEROGENEOUS COMBUSTION PROCESS IN AN HCCI ENGINE
OPTICAL CHARACTERIZATION OF FLUCTUATIONS AND LENGTH SCALES

Anders Hultqvist[1], Magnus Christensen[2], Bengt Johansson[2]
1 Scania CV AB 2 Lund Institute of Technology

ABSTRACT

An experimental study of the Homogeneous Charge Compression Ignition (HCCI) combustion process has been conducted by using chemiluminescence imaging. The major intent was to characterize the flame structure and its length scales and fluctuations. To achieve this, time resolved images of the naturally emitted light were taken in a port fuel injected heavy-duty engine equipped with a Bowditch type extended piston. All measurements were made under the same operating conditions and a mixture of 50% n-heptane and 50% iso-octane was used as fuel. The effect of port fuel injection was compared with that of fuel injection about 1.5 m upstream of the inlet. The compression ratio was set to 12:1.

It was established that cool flames exist at about 20 crank angles before top dead center. They appear with a faint structure, evenly distributed throughout the combustion chamber. The main heat release peaks around top dead center. Looking at a macro scale, we find that the charge starts to burn simultaneously at arbitrary points in the combustion chamber. Ignition has a preference to occur between the intake valve and the intake valve along the cylinder liner. Combustion then progresses radially towards the center of the cylinder. This effect is probably due to fuel droplets entering the combustion chamber. Most likely, these droplets are then, due to the swirling motion, transported towards the outer radii of the cylinder, creating richer areas where ignition can take place under more favorable conditions. Integral length scales of the size of the chemiluminescence structures of the HCCI combustion process were obtained. It was observed that the size increases with increasing rate of heat release until peak rate of heat release is reached. Then the structures become smaller and disappear as the charge burns out. The large spatial cycle-to-cycle variations of the combustion process were evaluated by the standard deviation in each image. Early in the cycle, the standard deviation is on the same order as the mean intensity but as combustion progresses, the standard deviation decreases to about half the mean value.

INTRODUCTION

In a Homogeneous Charge Compression Ignition (HCCI) engine, the fuel and air are premixed to create a homogeneous charge. During the compression stroke, the charge is heated to obtain auto-ignition close to TDC. The cycle-to-cycle variations of the HCCI combustion process are very small [1,2] since combustion initiation occurs in many places simultaneously [3]. Since ignition occurs at multiple points, the combustion rate becomes very high and therefore highly diluted mixtures or exhaust gas recirculation (EGR) have to be used in order to limit the rate of combustion [4]. With HCCI, there is no direct means to control the start of combustion as the ignition process relies on spontaneous auto-ignition. By adjusting the remote operating parameters e.g. inlet temperature or EGR rate, ignition timing can be controlled [5,6]. The major advantages of HCCI compared to the diesel engine are low NOx emissions and, depending on the fuel, virtually no soot [7]. The benefit of HCCI compared to the Spark Ignition (SI) engine is the much higher part load efficiency [8].

The toughest challenge is controlling the ignition timing over a wide load and speed range [9,10]. Another is to obtain an acceptable power density. The power density is limited by combustion noise and high peak pressures. At low loads, the rather high emissions of unburned hydrocarbons in combination with low exhaust temperatures present an additional challenge [11].

By using unthrottled HCCI combustion at part load in SI engines, efficiency can be improved by 40-100%. Emerging technologies for variable valve timing and the intent of using these as a means of controlling the combustion timing in HCCI engines [12], may well lead the way for a potential application of the HCCI combustion mode in practice.

The HCCI combustion process has previously, to some extent, been studied through different imaging techniques by Richter et al. [13,14] and other researchers [15,16,17]. The visualization of the combustion process has proven a valuable tool in understanding the combustion process. Experiments with chemiluminescence imaging [18] and planar induced fluorescence [19] indicate that the combustion process is delayed along the cool combustion chamber walls. However, combustion-to-wall interaction was not found to be the main mechanism responsible for the emissions of unburned hydrocarbons.

In the present experiment a truck-sized engine, with two different types of port fuel injection systems, has been studied. The main purpose of this work is to characterize the flame structure and its length scales and fluctuations. To achieve this, time-resolved images of the naturally emitted light has been taken at different crank angle positions. These images were then processed to give an estimate of the size of the chemiluminescence structures and of how much the size varies from cycle to cycle.

EXPERIMENTAL SETUP

ENGINE

A Scania D12 single cylinder engine was equipped with a Bowditch type optical access to enable imaging of most of the combustion chamber. To keep the vibrations from the oscillating masses on a low level, the engine was equipped with two pairs of counter rotating balancing shafts, contained in the oil sump. The engine and the dynamometer were rigidly joined with a floating framework, so as not to transfer any vibrations to the surroundings. On the intake and exhaust side, pressure equalization vessels were fitted. A piezoelectric pressure transducer was used to record the cylinder pressure. In Figure 1, the optical engine is displayed with its cylinder dropped down to its lower position for cleaning. The neat design, also used at the IFP and Sandia National Labs., with a "drop-down" liner, provides swift access to the inner glass surfaces for cleaning or inspection. A cleaning cycle is completed within five minutes, which can be compared to a conventional head disassembly, which requires about one hour.

Figure 1: Single cylinder Scania D12 engine

The compression ratio was set to 12:1 and a mixture of 50% n-heptane and 50% iso-octane was fed to the engine using separate injectors for each fuel. In the first case, an intake spacer was fitted to the cylinder head allowing the injectors to inject fuel in each port respectively. In the second case, the fuel was injected 1.5 m upstream of the inlet, promoting good mixing. In Table 1 the basic engine specifications are shown.

Table 1 : Geometric properties

Displaced Volume	*1951 cm^3*
Bore	*127 mm*
Stroke	*154 mm*
Connecting Rod	*255 mm*
Inlet Valve Seat Diameter	*2x 39.7 mm*
Exhaust Valve Seat Diameter	*2x 38 mm*
Valve Lift Exhaust	*15 mm*
Valve Lift Inlet	*15 mm*
Swirl Ratio	*1.7*
Compression Ratio	*12:1*
Combustion Chamber	*Pancake*

To initiate HCCI combustion with the selected compression ratio and fuel type, a higher inlet temperature is necessary. Therefore, the inlet air was preheated with an electrical heater. A low compression ratio was chosen to ensure that the quartz glass in the piston top would not break and consequently a low octane (RON=50) fuel was used. All imaging was done under the operating conditions shown in Table 2. As thermal considerations made steady-state running of the engine impossible, the running of the engine heated up the metal parts. Consequently, the inlet air temperature had to be governed to maintain proper combustion timing.

Table 2 : Operating conditions

Inlet Temperature	*60-90 °C*
Inlet Pressure	*1 bar (a)*
Engine Speed	*700 rpm*
N-heptane	*50%*
Iso-octane	*50%*
Lambda	*3.95-4.13*

OPTICAL ACCESS

An optical engine design of the Bowditch type was chosen since it provides optical access to a large part of the combustion chamber. The design allows 75% of the bore to be imaged. Another advantage is the ability of three-way optical access, which is crucial for laser diagnostics.

A sectioned view of the engine is shown in Figure 2. This engine consists of a supporting structure of U-beams(1), a "drop-down" cylinder liner(2), an upper part cylinder liner quartz ring(3), a quartz glass piston-crown(4) and a mirror(5) placed inside the piston extension(6) that bolts on a modified piston in a standard block. Great caution was taken when designing the elongated push rods, the dry-running piston rings and the parts made of quartz glass since these components were identified as critical to the engine operation. Having run the engine for six months, the team is confident that it works as intended.

In Figure 3, the optical engine is displayed from the exhaust side, fully assembled and ready to run. The piston is positioned in TDC in the figure and the piston crown can barely be seen through the quartz ring. Also visible in the figure is the skewed view through the piston extension and the piston crown, and it is possible to catch a glimpse of one of the exhaust valves. The cooling circuit brings cool water from the engine block into the lower part of the liner. On the upper opposite side of the liner, the coolant leaves the liner and enters the cylinder head. From the cylinder head, it then returns to the engine block.

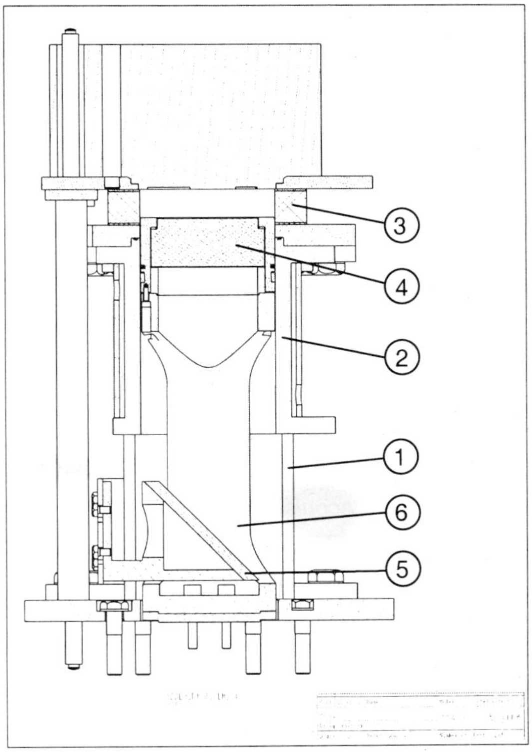

Figure 2 : Sectioned optical engine

Figure 3: Optical engine

IMAGE ACQUISITION SYSTEM

An intensified digital CCD camera was used for the imaging. ***Table 3*** summarizes the main features of the image acquisition system. A variety of gains, gate widths and apertures were tried out in order to determine the most favorable combination. As smearing due to charge motion was found to be a minor problem, a fairly long exposure time (200 microseconds) was chosen in favor of depth of field. By gating the intensifier synchronously with the engine, it was possible to acquire an image at a specific crank angle. Due to the limitation of only one image per cycle, the acquired data is processed to get an ensemble average at every crank angle and an estimate of the cycle-to-cycle variations. An f=100mm UV-lens was used throughout these experiments.

Table 3 : Image acquisition system

Camera Model	*Princeton Instruments (PI) ICCD-576/G/RB*
CCD-array	*576x384 pixels 22.5 x 22.5 μm size*
Dynamic Range	*14-bit*
Photocatode	*Red-blue UV-NIR*
Resolution	*120μm*
Camera Controller	*PI ST-130*
Trigger	*PI PG-200*

RESULTS

Data from the two test cases with different injection strategies, i.e. port fuel injection and upstream fuel injection, will be analyzed and compared in this section. Tests were performed under the same operating and inlet conditions. However, this resulted in retarded combustion timing for the case of upstream fuel injection. This indicates that a more homogenous charge will retard the combustion timing. Unfortunately, because of a malfunctioning heater, a complete data set for the upstream fuel injection case can not be presented. A few data points early in the cycle are missing. All images are presented in false colors since the acquisition system is not capable of separating colors. Earlier spectroscopic investigations [18] have shown that the chemiluminescence ranges from 280 nm to above 470 nm, so in its actual colors, the chemiluminescence would be bluish.

The analysis is focused on the main heat release since the cool-flame structures are very small, homogeneous, or undetectable with the present acquisition system. Imaging of the chemiluminescence is equivalent to a line of sight integration of the chemiluminescence through the height of the combustion chamber, which is an obstacle when estimating sizes in a plane. The line of sight integration leads to an overestimate rather than an underestimate of the size of the chemiluminescence structure, since overlapping structures along the line of sight are perceived as one. When looking at the results, this fact should be kept in mind.

In Figure 4 and Figure 5, the combustion chamber and the combustion process are viewed from underneath, through the elongated piston and the mirror. The images show 20 different cycles at maximum intensity of the combustion process in the port fuel injection and the upstream fuel injection case respectively. From these raw data, scaled to fit the color map, it is clear that there is a difference in combustion homogeneity between port fuel injection and upstream fuel injection. In the case of port fuel injection, the combustion has a preference to occur along the outer radii of the cylinder, whereas the fuel injected upstream burns simultaneously throughout the entire combustion chamber. The size of the structures also differs significantly. By just looking at the structures and keeping in mind that the valves are about 40 mm in diameter, one can estimate the average size, in the port injection case, to be on the order of 10-12 mm. Regarding the upstream case, the average size is estimated to be slightly less, say, 7-9 mm. In both cases, the valves are visible since they are made of stainless steel, thus having higher reflectance than the cylinder head made of cast iron. In the average images shown in Figure 10 and Figure 11, this is compensated by applying a “gain field”. The spatial cycle-to-cycle variations are also obvious. No cycles show any resemblance to one another on the smaller scales, which is an amazing fact given that the cycle-to-cycle variations in IMEP are so small. This is evidence of the combustion process being stochastic regarding spatial variations, but having a stable, expected, progress when integrating over the entire combustion chamber volume.

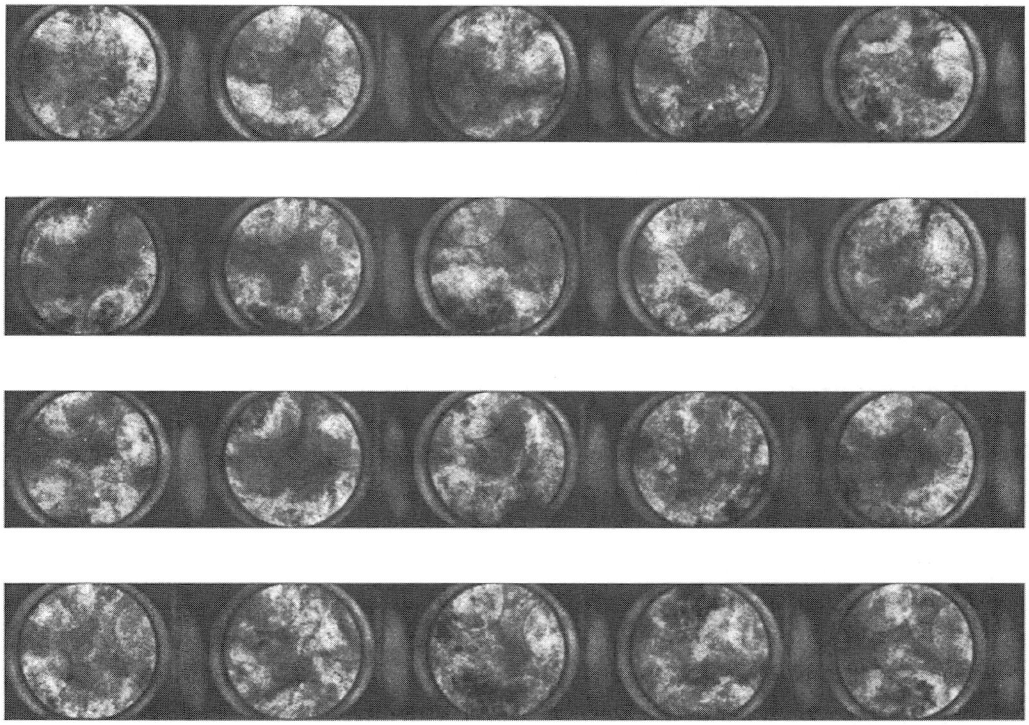

Figure 4: Port fuel injection, 20 cycles at maximum intensity, CAD = -1

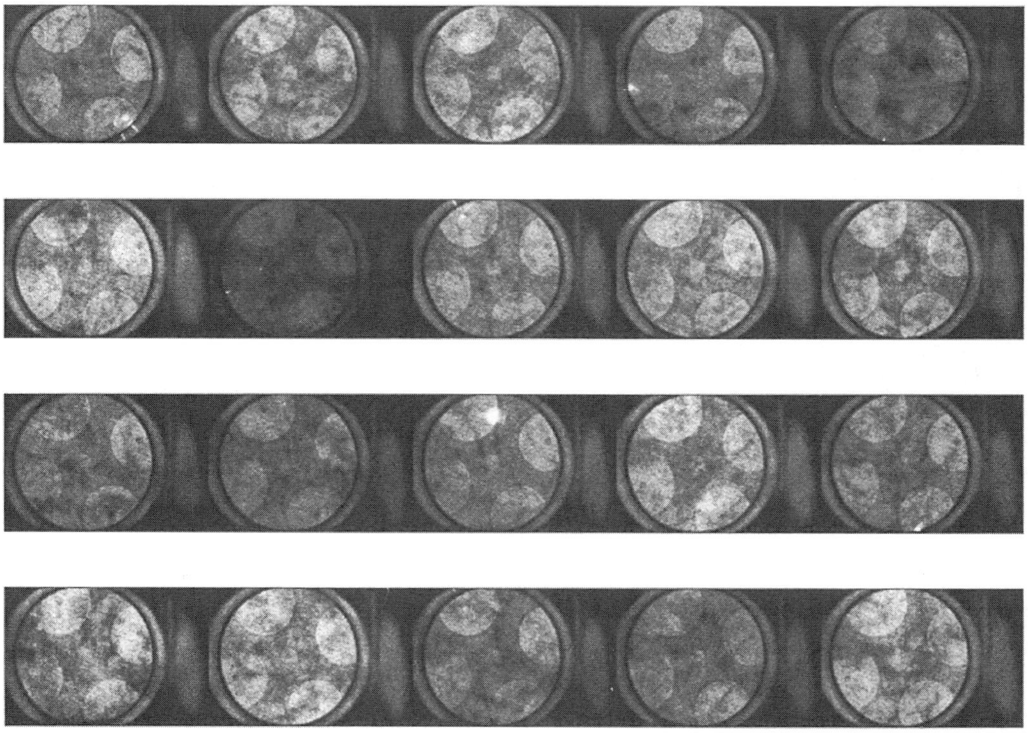

Figure 5: Upstream injection, 20 cycles at maximum intensity, CAD = 2

If we define a multi-dimensional matrix **I**, consisting of all the 20 images taken at each crank angle, every pixel intensity can be referred to by using the indices: x-pixel no, y-pixel no, N-frame no, CAD-crank angle. From the matrix **I**, the following can be defined:

Mean intensity of each image
$\mathbf{M}(N,CAD) = mean_2(\mathbf{I}(x,y,N,CAD))$

Average intensity
$\mathbf{MA}(CAD) = mean(\mathbf{M}(N,CAD))$

Standard deviation in each image
$\mathbf{S}(N,CAD) = std_2(\mathbf{I}(x,y,N,CAD))$

Average standard deviation
$\mathbf{SA}(CAD) = mean(\mathbf{S}(N,CAD))$

Standard deviation of standard deviations
$\mathbf{SS}(CAD) = std(\mathbf{S}(N,CAD))$

Here, *mean* and *std* refer to taking the mean and standard deviations respectively. The subscript 2 indicates that the operation is done along rows and columns.

Figure 6 and Figure 7 display the average standard deviation and the standard deviation of the standard deviations for the two test cases. Rates of heat release and average intensity are added for reference. In the case of port fuel injection, the average standard deviation is of the same size as the average intensity at the beginning of main heat release. At peak intensity the average standard deviation is about 41% of the average intensity. Further on, the deviation remains on the order of half the intensity. With upstream fuel injection the behavior is rather different. Throughout the combustion process the average standard deviation is below 50% of the average intensity, indicating less variation compared to the port fuel injection case. Around peak intensity, the average standard deviation is below 25% of the average intensity. The standard deviation of standard deviations indicates that the cycle-to-cycle variations are larger for the port injected case and that they are largest at the start of combustion.

Standard deviation, being the proper parameter describing a normal distribution, does not account for the probability densities found in the experiments. The averaged chemiluminescence probability density functions are presented in Figure 8 and Figure 9 to further illustrate the variations in combustion progress. The figures display the probability as a function of intensity. Since there is a very good correlation between chemiluminescence intensity and rate of heat release, one can interpret these figures as the distribution of heat release. In the plots, an intensity of 180 corresponds to the offset level of the intensified camera. This should be interpreted as zero rate of heat release.

Ideally, in a perfectly homogeneous charge the rate of heat release would be invariable. The ideal combustion process would then appear as a Dirâc spike, moving from zero rate of heat release towards maximum and back to zero as the charge burns out. This idealization is common amongst people involved with kinetics, but obviously, this is not the case in reality. The reactions and rate of heat release are distributed in space. This is a fact that should be taken should be considered when modeling the HCCI combustion process. In the case of port fuel injection and upstream fuel injection, the effect of charge homogeneity is obvious. A more homogeneous charge gives a narrower distribution of chemiluminescence intensity and it is possible to imagine a normal distribution moving along the intensity axis.

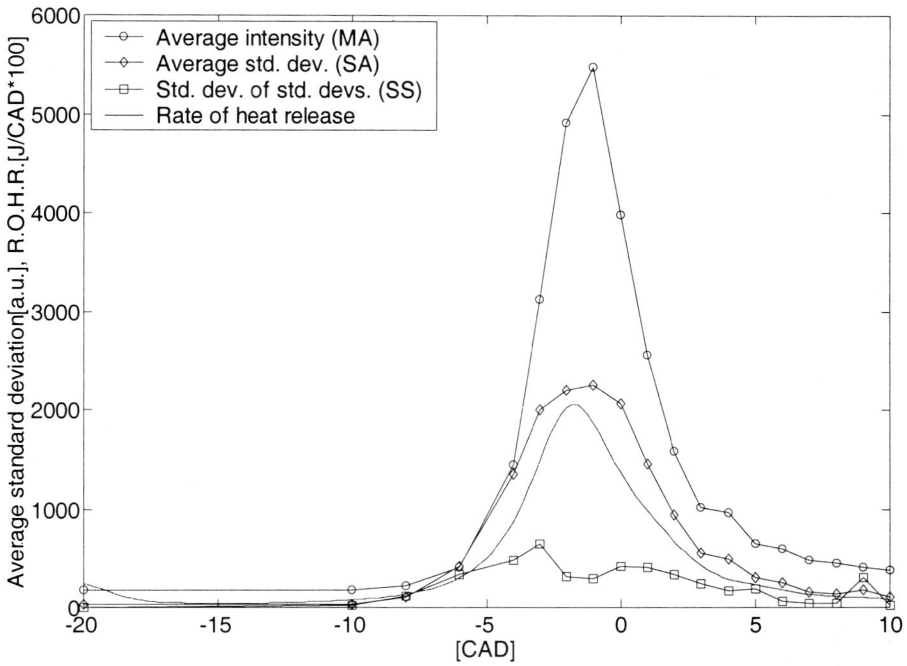

Figure 6: Port fuel injection, standard deviations and rate of heat release

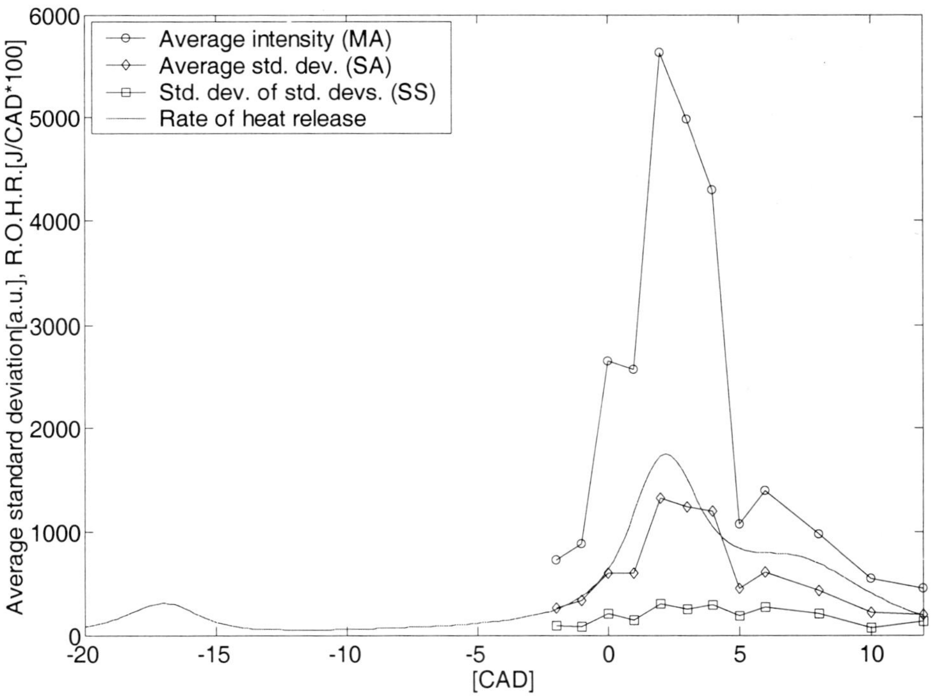

Figure 7: Upstream fuel injection, standard deviations and rate of heat release

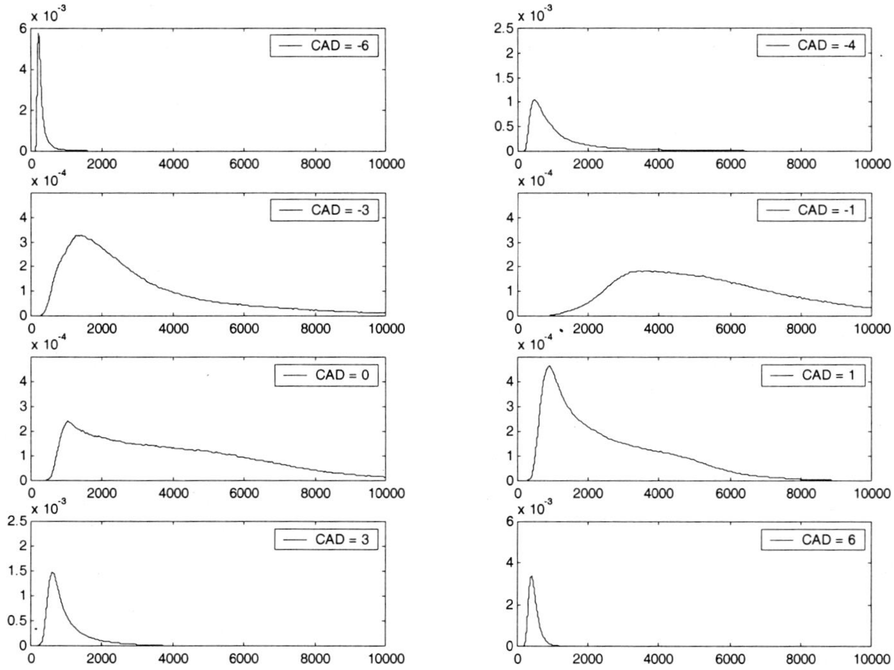

Figure 8: Port fuel injection, chemiluminescence probability density functions vs. intensity at different CADs

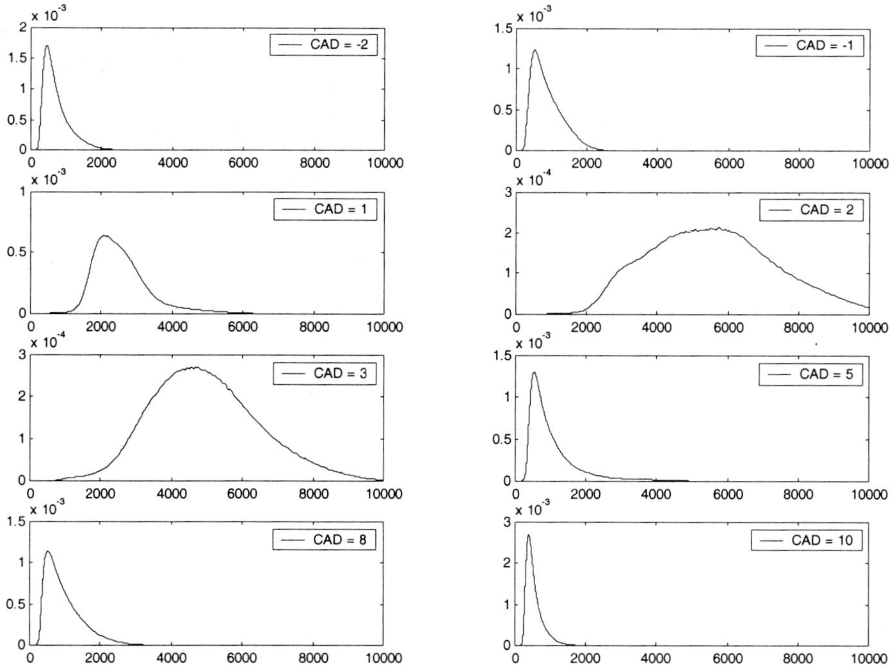

Figure 9: Upstream fuel injection, chemiluminescence probability density functions vs. intensity at different CADs

Average combustion development is displayed in Figure 10 and Figure 11. These images are ensemble averaged over 20 images and scaled to fit the color map. To suppress the effects of different reflectance of the valves and the cylinder head, a "gain field" normalized the mean images. The "gain field" was computed as the mean of all images. The "gain field" has a very smooth and credible appearance of how the reflectance might vary over the imaged area. In the images, the valves to the left are intake valves and the swirl motion is counter clockwise. Figure 10 shows that ignition of port injected fuel preferably occurs in the lower part of the image just between the intake and exhaust valves. With the inlet port direction and the swirl motion in mind, it is probable that fuel entering the cylinder as liquid would hit the cylinder wall right here. As the charge is compressed and the liquid fuel is vaporized, favorable rich conditions for ignition are created and hence the mixture is more likely to ignite here. Further on, from CAD = -4 to CAD = 4, combustion progresses from the outer areas of the cylinder towards the center, which was quite unexpected. In prior investigations on another heavy-duty engine [18], the authors have observed that combustion progresses from the core of the bulk towards the combustion chamber walls. A plausible explanation to this unexpected behavior is that the swirl motion transports larger droplets towards the cylinder liner during the intake stroke, thus creating a richer area along the outer radii of the cylinder. Although the case with fuel injected upstream shows a similar preference of ignition, the combustion swiftly moves towards a homogenous progress.

Analogously with turbulence theory, where different length scales are defined to characterize the turbulent flow field, it is possible to define an integral length scale for the chemiluminescence structures. The integral length scale is a measure of the average size of the largest structures in a flow field or, in this case, chemiluminescence. The integral length scale is here defined by the integral of the two-dimensional correlation coefficient, from zero to infinite displacement. Of course, it is not possible to integrate towards infinity since the imaged area is limited. It is practical to abort when the correlation coefficient approaches zero, ***imax***. The two-dimensional correlation coefficient ***r*** is calculated by the following expression:

$$r = \frac{\sum_m \sum_n \left(A_{mn} - \bar{A}\right)\left(B_{mn} - \bar{B}\right)}{\sqrt{\left(\sum_m \sum_n \left(A_{mn} - \bar{A}\right)^2\right)\left(\sum_m \sum_n \left(B_{mn} - \bar{B}\right)^2\right)}} \qquad (1)$$

In the expression, **A** and **B** represents the original image and the displaced one respectively. The integral length scale ***Ic*** of the chemiluminescence structure is then calculated by a discrete expression:

$$I_C = \int_0^{\infty} r \cdot dx \cong \sum_{i=1}^{i=i\max} r \cdot \Delta x \qquad (2)$$

In Figure 12 and Figure 13, examples of the calculated correlation coefficients are displayed as a function of the displacement between the original image and the shifted one. The shift was only performed in one direction. Also included is the cumulative integral of the correlation that eventually reaches the integral length scale. In the figures, this can be between 6-7 mm. There are no striking differences in appearance, when comparing the port fuel injection case with the upstream fuel injection case. At small displacements, the correlation should be close to one. If this is not the case, the measurement resolution is inadequate. In these measurements, the port fuel injection case exhibits a better correlation at small displacements, however, the resolution is good enough.

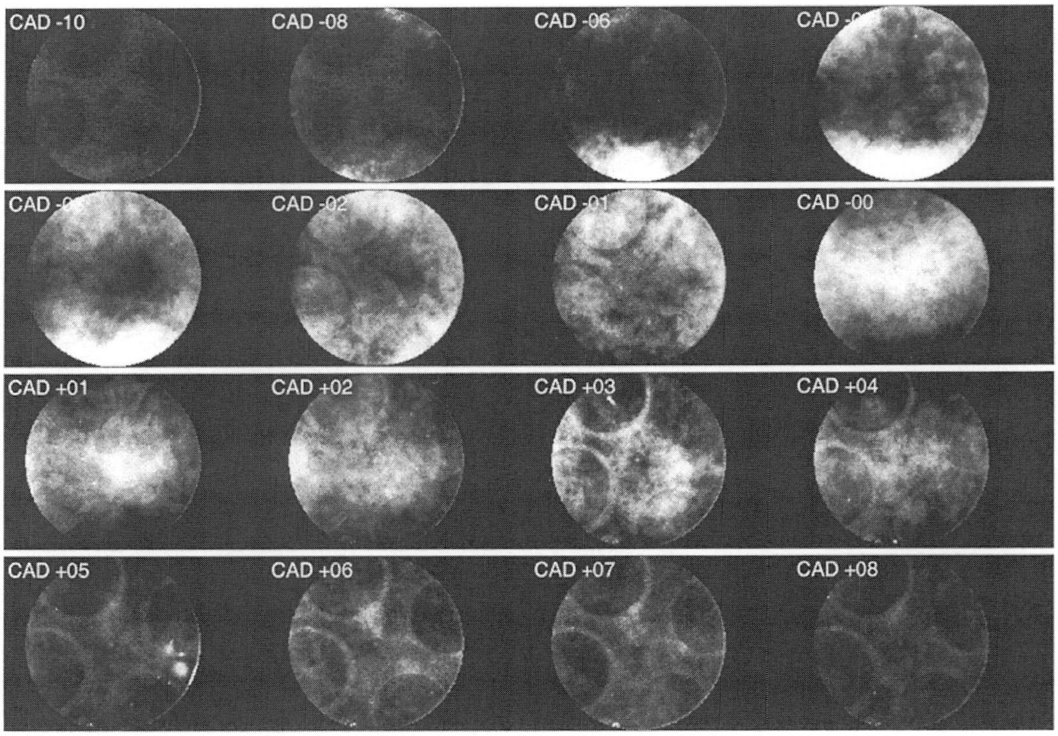

Figure 10: Port fuel injection, averaged combustion progress

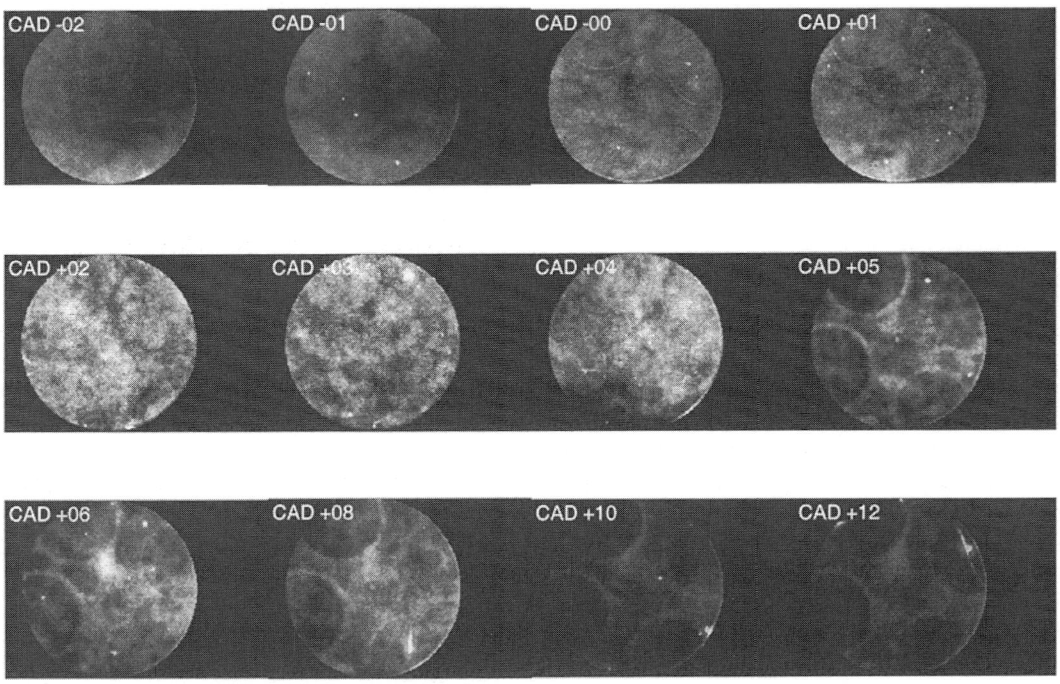

Figure 11: Upstream fuel injection, averaged combustion progress

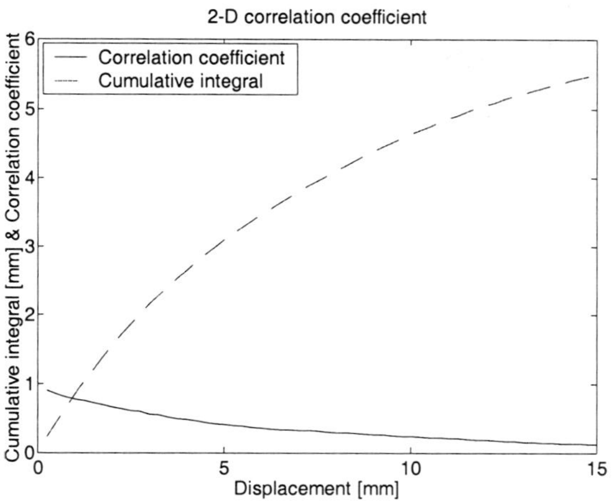

Figure 12: Port fuel injection, example of correlation coefficient

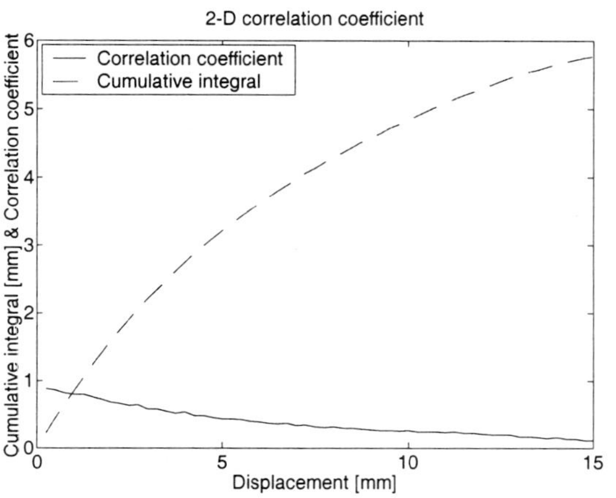

Figure 13: Upstream fuel injection, example of correlation coefficient

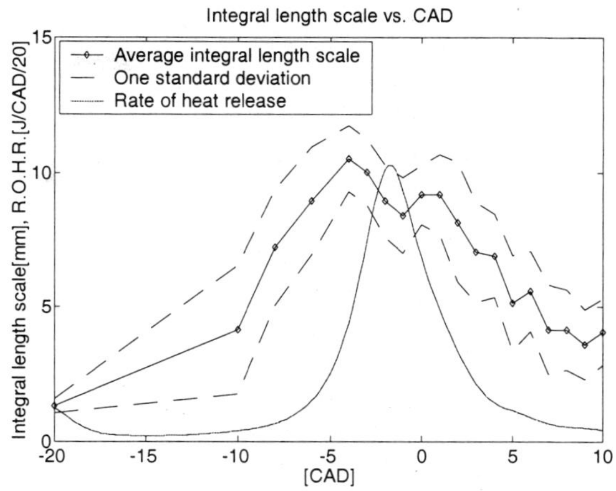

Figure 14: Port fuel injection, integral length scale

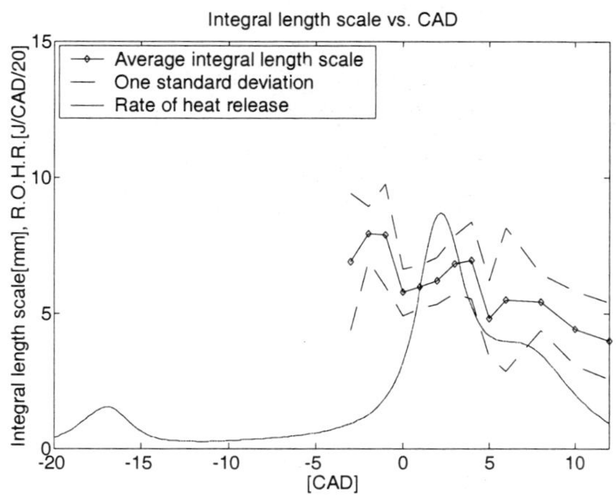

Figure 15: Upstream fuel injection, integral length scale

The development of the average integral length scale is showed in Figure 14 and Figure 15. Also included is the standard deviation of the integral length scale and the rate of heat release. From the case of port fuel injection in Figure 14, it can be concluded that the chemiluminescence structures start to grow early in the combustion progress. Their sizes tend to peak before the maximum rate of heat release at about 11 mm. During maximum rate of heat release, the size is fairly constant. This indicates that the reactions spread and that larger structures are formed before the major heat release.

SUMMARY AND CONCLUSIONS

An experimental study of the Homogeneous Charge Compression Ignition (HCCI) combustion process has been conducted by the use of chemiluminescence imaging. The major intent was to characterize the flame structure and its length scales and fluctuations. To achieve this, time resolved images of the naturally emitted light were taken in a port fuel injected heavy-duty engine. A Scania D12 single cylinder engine was equipped with a Bowditch type optical access to enable imaging of most of the combustion chamber. An intensified digital camera was used for the imaging. All imaging was performed under the same operating conditions and a mixture of 50% n-heptane and 50% iso-octane was used as fuel. The effect of port fuel injection is compared with that of fuel injection about 1.5 m upstream of the inlet. The compression ratio was set to 12:1.

It was established that cool flames exist about 20 crank angles before top dead center. They appear with a faint structure, evenly distributed throughout the combustion chamber. There was no luminosity recorded between the fading of the cool flames and the start of the main heat release. The main heat release peaks around top dead center. Looking at a macro scale, we find that the charge starts to burn simultaneously at arbitrary points in the combustion chamber. Ignition has a preference to occur between the intake valve and the intake valve along the cylinder liner. The combustion then progresses radially towards the center of the cylinder. This effect probably occurs due to fuel droplets entering the combustion chamber. These droplets then, as a result of the swirling motion, be transported towards the outer radii of the cylinder, creating richer areas where ignition can take place under more favorable conditions. Ensemble averaged images clearly show this ignition preference effect.

Through procedures similar to the defining of integral length scales in turbulence, the size of the chemiluminescence structures of the HCCI combustion process were obtained. It was observed that the size increases with increasing rate of heat release until peak rate of heat release is reached. Then the structures become smaller and disappear as the charge burns out. The size of the chemiluminescence structures peak at about 11 mm and this corresponds well to the constraining height of the combustion chamber around top dead center.

The large spatial cycle-to-cycle variations of the combustion process were evaluated by the standard deviation in each image. Early in the cycle, the standard deviation is on the same order as the mean intensity but as combustion progresses, the standard deviation decreases to about half the mean value. Standard deviation, being the proper parameter describing a normal distribution, does not account for the probability densities found in the experiments.

ACKNOWLEDGMENTS

The authors would like to extend their gratitude to Greger Juhlin at Scania CV AB and “Programrådet för Fordonsteknisk Forskning” (PFF) for sponsoring this investigation. They would also like to thank B. Andersson for his invaluable assistance and maintenance of the experimental apparatus. All of this work was conducted at the Lund Institute of Technology in Sweden.

REFERENCES

1. S: Onishi, S. Hong Jo, K. Shoda, P Do Jo, S. Kato: "Active Thermo-Atmosphere Combustion (ATAC) - A New Combustion Process for Internal Combustion Engines", SAE790501
2. R.H. Thring: "Homogeneous-Charge Compression-Ignition (HCCI) Engines", SAE892068
3. A. Hultqvist, M. Christensen, B. Johansson, A. Franke, M. Richter, M. Aldén: “A Study of the Homogeneous Charge Compression Ignition Combustion Process by Chemiluminescence Imaging”, SAE1999-01-3680
4. M. Christensen, B. Johansson: "Influence of Mixture Quality on Homogeneous Charge Compression Ignition", SAE9824541
5. T. Aoyama, Y. Hattori, J. Mizuta, Y. Sato: "An Experimental Study on a Premixed-Charge Compression Ignition Gasoline Engine, SAE960081
6. J-O. Olsson, O. Erlandsson, B. Johansson: " Experiments and Simulation of a Six-Cylinder Homogeneous Charge Compression Ignition (HCCI) Engine", SAE2000-01-2867
7. M. Christensen, A. Hultqvist, B. Johansson: “Demonstrating the Multi Fuel Capability of a Homogeneous Charge Compression Ignition Engine with Variable Compression Ratio",SAE1999-01-3679
8. M. Christensen, P. Einewall, B. Johansson: "Homogeneous Charge Compression Ignition (HCCI) Using Iso-octane, Ethanol and Natural Gas- A Comparison to Spark Ignition Operation" , SAE972874
9. J-O. Olsson, P. Tunestal, B. Johansson: " Closed-Loop Control of an HCCI Engine", SAE2001-01-1031
10. J-O. Olsson, P. Tunestal, G. Haraldsson, B. Johansson: "A Turbo Charged Dual Fuel HCCI Engine", SAE2001-01-1896
11. O. Erlandsson, B. Johansson, F. A. Silversand: "Hydrocarbon (HC) Reduction of Exhaust Gases from a Homogeneous Charge Compression Ignition (HCCI) Engine Using Different Catalytic Mesh-Coatings", SAE2000-01-1847
12. D. Law, J. Allen, D. Kemp, G. Kirkpatrick, T. Copland: “ Controlled Combustion in an IC-Engine with a Fully Variable Valve Train", SAE 2001 World Congress, March 2001, Detroit, MI, USA
13. M. Richter et al.:" Optical Diagnostics Applied to a Naturally Aspirated Homogeneous Charge Compression Ignition Engine", SAE 1999-01-3649, 1999
14. M. Richter, J. Engström, A. Franke, M. Aldén, A. Hultqvist and B. Johansson “The influence of charge inhomogeneity on the HCCI combustion process", SAE 2000-01-2868, SAE Fuel & Lube Fall Meeting, Baltimore, 2001.
15. H. Akagawa, T. Miyamoto, A. Harada, S. Sasaki, N. Shimazaki, T. Hashizume, K. Tsujimura:

"Appraoches to solve problems of the premixed lean diesel combustion", SAE1999-01-0183
16. Y. Iwabuchi, K. Kawai, T. Shoji, Y. Takeda: "Trial of a New Concept Diesel Combustion System - Premixed Compression-Ignited Combustion", SAE1999-01-0185
17. G. Andrews, K. Zaidi: " Partial Premixed Diesel Gaseous and Particulate Emissions", SAE1999-01-0838
18. A. Hultqvist et al.: " A Study of the Homogeneous Charge Compression Ignition Combustion Process by Chemiluminescence Imaging ", SAE 1999-01-3680, 1999
19. A. Hultqvist, U. Engdar, B. Johansson, J. Klingmann: " Reacting Boundary Layers in a Homogeneous Charge Compression Ignition (HCCI) Engine", SAE 2001 World Congress, March 2001, Detroit, MI, USA

CONTACT

The main author of this paper was awarded his M.Sc. in mechanical engineering at Lund Institute of Technology in 1996. He has been with Scania CV AB since then and is currently working on his Ph.D.

Lund Institute of Technology, Division of Heat and Power Engineering, Dep. Of Combustion Engines, P.O. Box 118, 221 00, Lund, Sweden

E-mail: anders.hultqvist@vok.lth.se

A New Generation of Engine Combustion Processes for the Future?
© P. Duret (Editor) and Editions Technip, Paris, 2001, pp. 69 - 76
27 rue Ginoux, 75015 Paris

FUEL INTRODUCTION STRATEGIES FOR PRE-MIXED COMPRESSION-IGNITION COMBUSTION

Rudolf H. Stanglmaier, Thomas W. Ryan III, Darius Mehta
Southwest Research Institute

ABSTRACT

A pre-mixed charge suitable for compression ignition combustion can be achieved by fuel injection (or fumigation) upstream of the intake valves or by direct injection of fuel into the combustion chamber. These strategies differ in the time available for mixing and in the degree of mixture homogeneity that is achieved at the time of combustion. The choice of fuel introduction strategy also has a profound effect on the required properties of the fuel and on the methods available for controlling the combustion phasing and heat release rate. This paper examines the use of port fuel injection (PFI) and direct fuel injection (DI) for HCCI combustion, as well as the relative benefits and disadvantages of each strategy. The arguments presented in this paper are drawn form the experiences of Southwest Research Institute at operating HCCI engines, as well as from the results reported by other research groups.

INTRODUCTION

Pre-mixed charge compression ignited combustion has emerged over the past five years as the most promising in-cylinder NO_x control technique for internal combustion engines. Some variations on this concept include homogeneous charge compression ignition (HCCI), controlled auto-ignition (CAI), and pre-mixed charged compression ignition (PCCI)*. Fundamentally, all of these methods rely on the low-temperature, distributed reaction of fuel-lean and/or highly diluted mixtures by auto-ignition. This combustion mode is advantageous in comparison to spark-ignition or Diesel combustion because the reduced combustion temperatures result in extremely low levels of NO and NO_2 production. Pre-mixed, or homogeneous, charge compression ignition combustion is also characterized by high heat release rates and reduced heat rejection to the engine coolant. Hence, when the combustion event is properly phased in relation to the engine cycle, this combustion mode can be conducive to high thermodynamic efficiencies. Nonetheless, the application of HCCI combustion to production engines is not without significant challenges. Some of the most notable difficulties include controlling the phasing of the combustion event relative to the engine cycle, and operating the engine under high loads.

Many variations on the concept of pre-mixed charge compression ignition are being pursued for application on motorcycles, passenger cars, medium and heavy-duty trucks, and even stationary power generation. One basic distinction between the various combustion concepts being developed can be drawn between those that are intended for full-time HCCI operation, and those that are designed for HCCI operation only over part of their load-speed range. Another fundamental distinction can be made according to the fuel introduction strategy employed, and this is the subject of the present paper. The choice of fuel introduction method guides many design constraints of an HCCI engine, and has a significant impact on the potential fuel efficiency and emissions of the engine. This paper discusses the main fuel introduction strategies and the relative benefits of each of them.

FUEL INTRODUCTION STRATEGIES FOR HCCI COMBUSTION

Pre-mixed charge engines can employ a variety of devices for fuel metering. HCCI engines have been reportedly operated with carburetors, gaseous fuel fumigators, fuel injectors located upstream of the intake valves (anywhere from the intake ports to one meter upstream of the valves), and medium to high-pressure fuel injectors in the combustion chamber.

* For convenience, the term HCCI will be used throughout this paper to cover CAI and PCCI combustion.

While each of these devices has individual nuances, the principal distinction is for fuel introduction upstream of the intake valves (referred to hereon as Port Fuel Injection or PFI) and fuel introduction directly to the combustion chamber (referred to hereon as Direct Injection or DI). This is illustrated schematically in Figure 1. The three fuel introduction strategies discussed in this paper are: Port Fuel Injection (PFI), Early-cycle direct injection (Early-DI), and Late-cycle direct injection (Late-DI). Early and Late DI differ on when the fuel is injected into the combustion chamber. Early-DI is performed into a relatively low-density air charge during the intake stroke or early during the compression stroke, whereas Late-DI is performed into a high-density air charge shortly before combustion or near TDC.

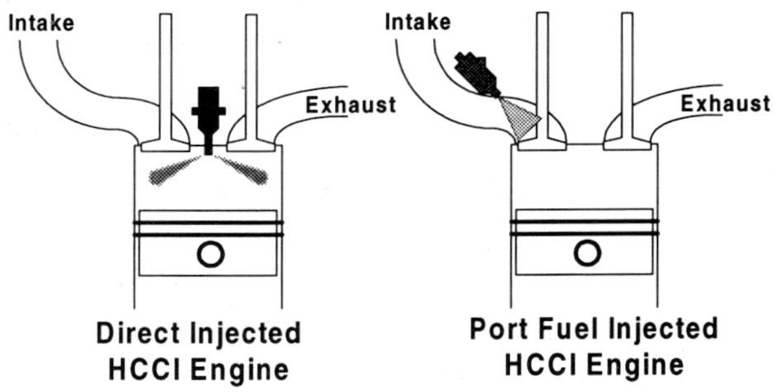

Figure 1. Schematic Illustration of Direct Injection (DI) and Port Fuel Injection (PFI) HCCI Engines.

Although not a requirement, port fuel injection is typically performed synchronously with the engine cycle (This is not the case with carburetors or fumigators). However, the fuel is inducted into the combustion chamber only when the intake valves open and the fuel-air mixture is carried into the cylinder. As a result, a relatively long time is available for injecting fuel into the port (nominally 720 CAD). This time interval is also available for heat transfer and vaporization prior to induction into the cylinder. Direct Injection, on the other hand, must be performed synchronously with the engine cycle, and during a shorter time interval. Early-DI is typically performed within 180 CAD, roughly corresponding with the duration of the intake stroke. For late-cycle DI engines, the entire fuel mass must be introduced to the combustion chamber within the ignition delay (typically 1-2 ms, or 7-14 CAD at 1200 RPM), and the fuel must vaporize and mix with the air charge during this time. These three fuel introduction strategies require fuel injection equipment with very different characteristics, as summarized in Table 1.

	Port Fuel Injection	**Early-cycle DI**	**Late-cycle DI**
Time available for fuel injection	Up to 720 CAD	Up to ~ 220 CAD	~ 5 to 30 CAD
Typical injection pressure	2 to 5 bar	45 to 100 bar	650 to 2500 bar
Droplet size range	50 to 300 μm SMD	15to 30 μm SMD	15 to 30 μm SMD
Multiple injection capability	Not useful	Yes	Yes
Relative cost	Low	Medium	High

Table 1. Characteristics of Fuel Systems Used in HCCI Engines.

COMBUSTION PHASING CONTROL

Unlike in Diesel or Spark-Ignited (SI) engines, a direct method for controlling the start of combustion is not available in HCCI engines. Instead, the start of combustion is established by the low-temperature auto-ignition chemistry of the local air-fuel mixture. Low temperature auto-ignition of a fuel-oxidizer mixture is influenced by the properties of the mixture and by time-temperature history to which it is exposed. Correspondingly, there are two fundamental approaches to controlling the combustion phasing: (1) by adjusting the mixture reactivity, and (2) by adjusting the time-temperature history to which it is exposed. This is illustrated in Figure 2, along with various specific ways of implementing the two control approaches.

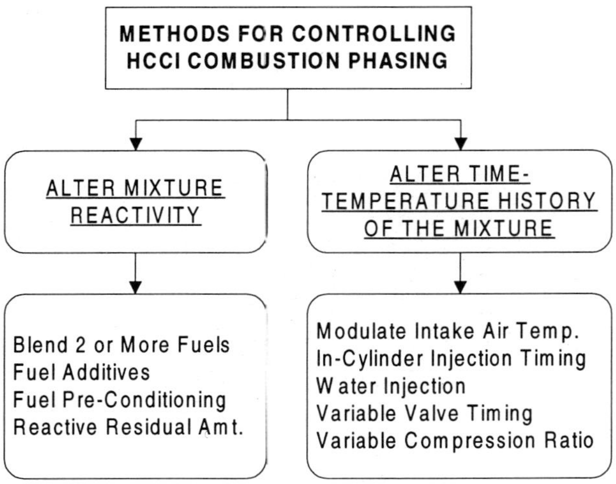

Figure 2. Methods for Controlling HCCI Combustion Phasing.

Direct fuel injection into the combustion chamber of a pre-mixed charge compression ignition engine allows the possibility of controlling the start of reaction through the injection timing. However, this has only been achieved successfully when the injection event occurs late in the cycle (Late-DI). Early-DI is less effective at controlling the start of combustion because the fuel is injected when the cylinder temperatures are relatively low and the charge-cooling effect associated with the vaporization process can alter the time-temperature history of the mixture (Takeda et al., 1996; Nakagome et al., 1997). In some cases, retarded injection timings can result in advanced start of combustion, and vice-versa. An additional difficulty with the Early-DI strategy is that it can result in large fractions of the fuel impinging on the cylinder liner and piston-top, and in correspondingly high levels of HC emissions and oil dilution. From the standpoint of combustion phasing control, Late-DI is clearly advantageous over Early-DI and PFI, as it requires no additional method for controlling the start of combustion.

One example of HCCI combustion phasing control through injection timing is shown below. A single-cylinder Cummins ISB engine was modified at SwRI to accommodate a gasoline direct fuel injector and a spark plug (not used during HCCI operation), as shown in Figures 3 and 4. Some characteristics of the engine and fuel injection system are listed in Table 2. Figure 5 shows a family of cylinder pressure traces obtained by operation of this engine on various injection timings. This plot illustrates that injection timing was effectively used to control the start of combustion in a Late-DI HCCI configuration.

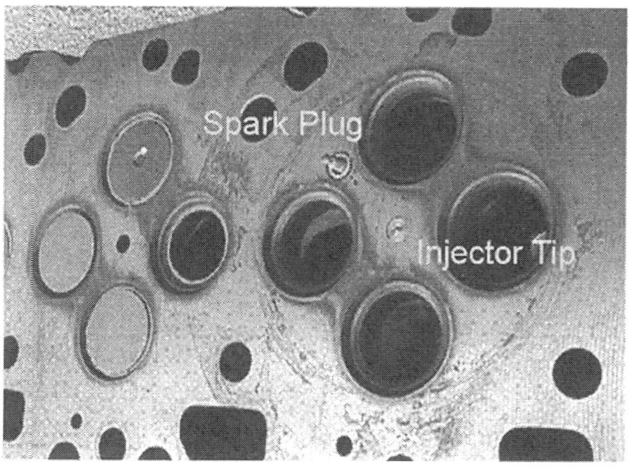

Figure 3. Single-cylinder Cummins ISB engine modified to operate as a DI-HCCI engine.

The time interval between start of injection and start of combustion (ignition delay) can be affected by changing the compression ratio, the intake air temperature, or the auto-ignition characteristics of the fuel.

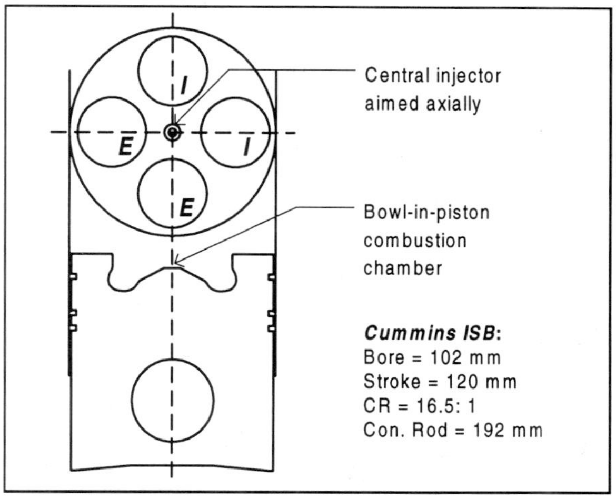

Figure 4. DI-HCCI engine combustion chamber.

Bore x Stroke	102 x 120 (mm)
Conn. Rod Length	192 (mm)
Compression Ratio	16.5:1
Combustion Chamber Shape	Bowl in Piston
Valves per Cylinder	4
Injector Type:	Hollow-cone, outward-opening pintle, swirl
Injector Location	Central, Aimed Axially
Injector Cone Angle	60 degrees
Fuel Supply Pressure	10 (MPa)
Drop Size (SMD)	15-20 μm

Table 2. Characteristics of single-cylinder DI-HCCI engine of Figure 3.

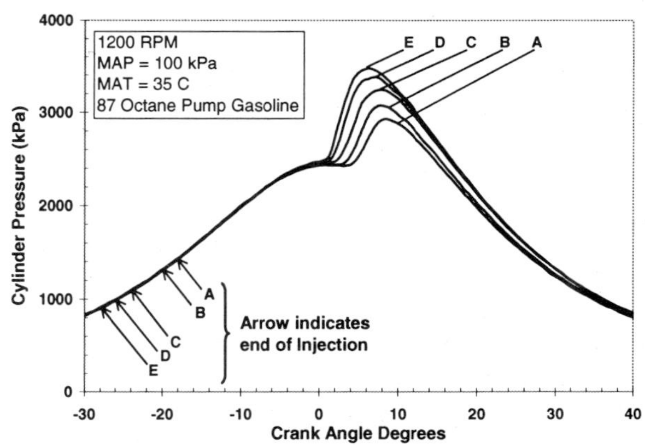

Figure 5. Cylinder pressure traces from a Late-DI engine, illustrating combustion phasing control. Note that the injection event is completed before the start of combustion.

Other methods for controlling the start of HCCI combustion that are well documented include fuel-blending (e.g. Olsson et al., 2001; Stanglmaier et al., 2001), variable valve actuation (e.g. Law et al., 2001), and variable compression ratio (e.g. Gray and Ryan, 1997; Christensen et al., 1999). These methods have all been reported with PFI, but could potentially be used with DI as well.

MIXTURE HOMOGENEITY

Port-fuel injection takes advantage of the turbulence generated by the intake charge as it flows across the intake valves to achieve a very well-mixed charge within the cylinder*. Early-DI typically also produces a well-mixed charge, provided that the in-cylinder charge motion is designed to disperse the injected fuel. Late-DI, on the other hand, can result in a highly inhomogeneous mixture at the time of combustion. The degree of inhomogeneity of the mixture at the time of combustion depends on how quickly the fuel can be injected, vaporized, and mixed with the surrounding air charge in the short time between the start of injection and the beginning of combustion. Mixture homogeneity can have a significant impact on the emissions characteristics, as discussed next.

EMISSIONS VS. MIXTURE HOMOGENEITY

NO_x Emissions - The low NO_x emissions obtained through HCCI combustion are a result of the distributed low-temperature reactions that occur when the fuel-lean mixture is auto-ignited. As the air-fuel mixture homogeneity is decreased, and the local air-fuel ratio in some regions becomes significantly lower than the global air-fuel ratio, the NO_x emissions increase. Thus, from a NO_x standpoint, a high degree of homogeneity is desirable. However, very low NO_x emissions can still be obtained by auto-ignition of relatively inhomogeneous mixtures, as long as the local combustion temperatures remain low.

HC Emissions - Two important mechanisms for HC emissions from homogeneous charge engines are crevices and liquid fuel effects. HC emissions are generated when fuel that is packed inside the crevices of the combustion chamber escapes combustion, and is not fully oxidized during the expansion stroke (Cheng et al., 1993). Liquid fuel impingement onto the piston-top and cylinder liner has also been shown to dramatically increase HC emissions from DI engines (Stanglmaier et al., 1999). Crevices can have a very strong effect on the HC emissions from HCCI engines, since the lean air-fuel ratios they operate on result in very low in-cylinder temperatures after combustion. Hence, any fuel that escapes combustion has a very low chance of undergoing complete post-combustion oxidation. Liquid fuel impingement on the walls of the combustion chamber is most likely to occur with direct fuel injection. In fact, some wall wetting is virtually unavoidable for Early-DI of liquid fuels. Wall wetting can be minimized or avoided completely with Late-DI, but this requires a careful balancing of the injector characteristics, in-cylinder air flow, and fuel properties. In-cylinder wall wetting is usually avoided with PFI, as the fuel is fully vaporized before being inducted into the cylinder.

Port fuel injected and Early-DI HCCI engines are probably destined to have relatively high engine-out HC emissions because they pack a homogeneous air-fuel charge into the top-land crevice. Early-DI engines have the added contribution to HC emissions from wall-wetting. Thus, from an HC emissions standpoint, Late-DI HCCI engines have the potential for producing the lowest levels of unburned and partially-burned HCs, provided that wall-wetting can be minimized or avoided.

Soot and PM Emissions - Soot and particulate matter (PM) are generated in regions where overly rich mixtures are combusted, such as diffusion flames. Within an engine, this occurs in the immediate vicinity of unmixed fuel jets, or around liquid fuel drops or pools. Port fuel injected engines generally achieve full vaporization of the fuel and thorough mixing, so they emit negligible levels of soot and particulate matter. Direct-injected engines, however, can produce high levels of soot and PM if the fuel is undermixed and/or if liquid fuel pools are generated within the cylinder. Work at SwRI has shown that in order to eliminate the possibility of generating high soot emissions from HCCI engines, it is essential to vaporize all of the fuel prior to the beginning of combustion. Hence, HCCI engines with PFI have an inherent advantage in PM emissions over their DI counterparts.

* It is impossible to achieve a completely "homogeneous" mixture, as temperature and composition gradients will always be present. However, the magnitude of these gradients achieved with PFI is relatively small.

NO_x vs. HC Emissions in DI Engines - As discussed, thorough mixing in HCCI engines is advantageous for low NOx production, but crevice packing and wall-wetting is detrimental to HC emissions. In DI engines, this causes a competing effect between these two pollutants, when optimizing the injection timing. Figure 6 shows the NO_x and HC emissions as the injection is varied in a DI HCCI engine. Early (advanced) injection timings produce very low NO_x emissions because the charge is thoroughly mixed and localized combustion temperatures are kept low. However, Early injection results in higher HC emissions due to increased crevice packing and wall wetting. Late (retarded) injection timings have significantly reduced HC emissions by reducing the amount of fuel that is deposited on the walls and packed into the crevices. However, late injection results in a more inhomogeneous mixture and much higher NO_x levels.

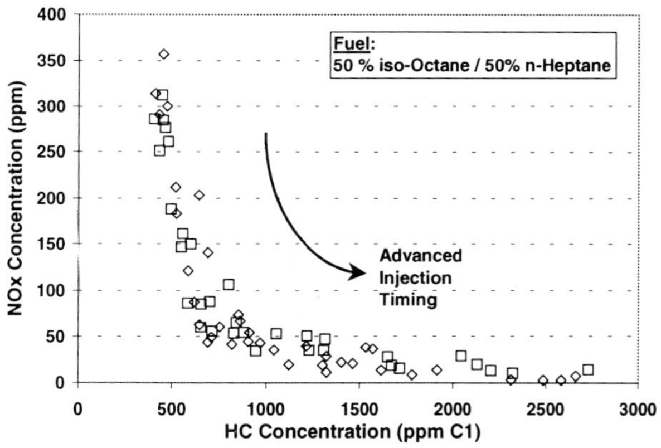

Figure 6. NO_x vs. HC emissions in a Late-DI engine, as the injection timing is varied.

The NO_x-HC tradeoff curve shown in Figure 6 is determined by the mixing rate of the fuel within the combustion chamber. An improvement in this tradeoff curve can be obtained by intensifying the fuel-air mixing process within the cylinder such that retarded injection timings can be used without generating overly rich regions. This may be achieved with different fuel injector designs and/or enhanced in-cylinder air flow.

Figure 7 provides a relative comparison of the exhaust emissions concentration of HC and NO_x from a PFI-HCCI engine and the DI-HCCI engine of Figure 6. The PFI engine was fueled on natural gas and Fisher-Tropsh naphtha (using fuel-blending for combustion phasing control). This engine and its emissions have been described in detail in other publications (Stanglmaier et al., 2001, and Gale et al., 2001). The PFI and DI HCCI engines were very different and used very different fuels, so this comparison should only be used on relative terms. Nonetheless, this plot illustrates that direct injection with retarded timings (Late-DI) can be used to reduce the HC emissions from an HCCI engine by eliminating the effect of the top-land crevice. Although not specifically illustrated in Figure 7, the injection timing of the PFI engine has no effect on the emissions of the engine.

Similarly, Figure 8 shows a comparison of the HC and CO concentration from the PFI-HCCI and DI-HCCI engines. Again, this plot is only meant to provide a general comparison of the characteristics of these two engines. However, it can be observed that the retarded DI injection timings that are conducive to low HC emissions are also conducive to low CO emissions. There is no clear correlation between the HC and CO emissions from the PFI engine.

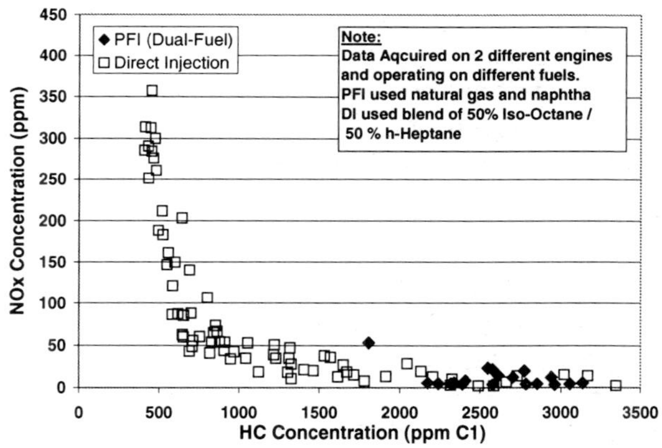

Figure 7. Comparison of HC and NO_x emissions from PFI-HCCI and DI-HCCI engines.

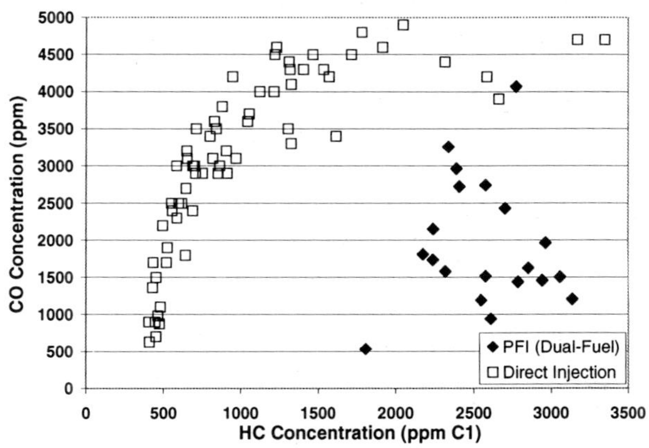

Figure 8. Comparison of HC and CO emissions from PFI-HCCI and DI-HCCI engines.

HEAT-RELEASE RATE VS. MIXTURE HOMOGENEITY

A significant difficulty facing HCCI engines is that the heat release rates are very high, and generally increase as the air-fuel ratio is reduced. This can lead to detonation or knock, and ultimately limits the power obtainable from such engines. The heat release rate of an HCCI engine can be reduced somewhat by the addition of EGR and by utilizing fuels with inherently slower combustion speeds. One additional recourse for reducing the heat release rate of HCCI engines is to reduce the mixture homogeneity. It has been shown that stratification of the charge can be used to moderate heat release (e.g. Morimoto et al., 2001). Charge stratification can be accomplished by introducing unmixed EGR, or by limiting the mixing of fuel with the remainder of the charge. This suggests that late-cycle DI may be used to produce a stratified charge (yet pre-mixed to avoid diffusion-limited combustion) to moderate the heat release rate and allow higher BMEP levels from an HCCI engine.

FUEL PROPERTIES

Fuels for HCCI operation require a different set of properties than those for spark-ignited or Diesel combustion. Furthermore, different properties are required for HCCI operation of a PFI configuration than a DI engine. Two properties that are very important for HCCI engines are volatility and reactivity (or propensity for low temperature auto-ignition)[†]. The volatility range must be matched to the fuel injection strategy to achieve proper mixture preparation, while the reactivity of the fuel (or fuels) must be selected to achieve combustion that is properly phased with the engine cycle.

As indicated previously, a pre-requisite for effective operation of an HCCI engine is that a great majority of the fuel must be vaporized before the start of combustion. Incomplete fuel vaporization can lead to incomplete combustion and increased levels of HC, CO, and PM emissions. To achieve this in a PFI configuration requires a fuel that is sufficiently volatile to vaporize under intake manifold conditions. It is possible for PFI to generate small fuel droplets that follow the air-stream into the combustion chamber and continue to vaporize within the cylinder. However, liquid fuel in the cylinder may cause wall-wetting and lead to increased HC and PM emissions. Fuel vaporization in the port is made easier when the fuel is injected onto the back of the hot intake valves and allowed to vaporize over a large portion of the engine cycle before being inducted into the cylinder (eg. Stanglmaier et al., 1997). Note that high intake manifold pressures and reduced intake air temperatures are detrimental to fuel vaporization, although they are desirable for improving volumetric efficiency. Early-DI engines impose perhaps the most severe requirement on fuel vaporization. In Early-DI engines, the fuel is injected into a relatively low-temperature and low-density environment, yet the fuel droplets are required to vaporize before they impinge on the piston-top or cylinder liner. Hence, all of the enthalpy of vaporization must be extracted from the surrounding air charge by convection. Effective mixture preparation through early-cycle DI requires very fine sprays and very volatile fuels. Late-cycle DI is most similar to Diesel combustion, except all of the fuel is injected prior to combustion. However, during Late-DI operation, the fuel is injected into a high-temperature and high-density air charge, which helps control the spray penetration and enhances the vaporization process. In general, it is possible to use lower volatility fuels with Late-DI than with Early-DI or PFI, unless elevated intake air temperatures are employed.

The start of combustion in an HCCI engine is determined by the low temperature auto-ignition properties of the fuel (or fuel mixture). Fuel that is introduced by PFI or Early-DI will undergo the entire compression stroke and thus follow the time-temperature curve prescribed by the engine speed and compression ratio. Thus, a fuel or combination of fuels must be used that produces auto-ignition in the right range of timings for a given engine geometric design. It is possible to tailor a fuel for a specific compression ratio or to design the compression ratio to suit a certain fuel. However, the most promising results suggest PFI and Early-DI engines should use fuels that have relatively low reactivity (e.g. Christensen, et al., 1999). In Late-DI engines, all of the fuel must be injected within the ignition delay period, but the fuel is only exposed to the high temperature environment for a relatively short time. For compression ratios in the range used in Diesel engines (16-20:1), this requires a fuel that is generally more reactive than gasoline (lower Octane number), but less reactive than Diesel fuel (lower Cetane number). Compared to PFI and Early-DI, Late-DI may require fuels with a higher range of reactivity.

[†] Note that the Octane and Cetane scales are not directly applicable to HCCI fuels, and should be used with care in this context.

Late-cycle DI engines may have an additional fuel property requirement. The injection equipment required to produce the high injection pressures used in Late-DI may require fuels with substantial lubricity characteristics. The low and medium pressure fuel injection equipment used in PFI and Early-DI engines usually has no lubricity requirement. A summary of relative fuel property requirements for the three injection strategies is presented in Table 3.

	Port Fuel Injection	Early-cycle DI	Late-cycle DI
End distillation point	~ 200 - 220 °C or lower	Best if less than ~ 160°C	Can use 300 - 350 °C
Relative reactivity	Low to medium	Low to medium	Medium to high
Lubricity requirement?	None	Low	Yes, depending on system

Table 3. Summary of desired fuel properties for HCCI engines.

CONCLUSIONS AND SUMMARY

It is possible to operate pre-mixed charge compression ignition engines with port fuel injection or direct fuel injection into the combustion chamber. The choice of fuel introduction strategy affects the options that are available for controlling the combustion phasing, the fuel properties that are required, and the emissions characteristics of the engine. The attributes of the main strategies are summarized below:

Early-Cycle DI - Early-DI is the least attractive of the three fuel introduction strategies discussed in this paper. It provides no benefit in controlling the start of combustion, and it requires fuel injection equipment that is relatively costly. In addition, it results in high CO and HC emissions due to full packing of the crevices, and the highest likelihood of promoting fuel wall wetting. Early-DI engines require fuels that are very volatile and with a relatively low reactivity.

Port Fuel Injection - PFI can probably achieve the lowest NO_x and PM emissions by producing the most homogeneous mixtures. However, by the same virtue, PFI engines will also produce relatively high HC and CO emissions. PFI requires the lowest complexity and cost components for fuel injection. However, it requires some external method for controlling the combustion phasing. The cost of a variable valve actuation (VVA), variable compression ratio (VCR), or secondary fuel system (for fuel blending) may reduce the cost benefit over DI. PFI engines require fuels that are very volatile and with a relatively low reactivity.

Late-Cycle DI - Late DI requires the most sophisticated and costly injection equipment, as fuel must be injected in a very short time, kept from wetting the walls, and mixed with the air charge very rapidly. PM emissions may be the most difficult pollutant to mitigate with Late-DI. However, HCCI engines that use Late-DI may simultaneously achieve low emissions of HC, CO, and NO_X. In addition, Late-DI can help manage the heat release rate of HCCI engines by controlling the mixture homogeneity at the start of combustion. Late-DI engines can use fuels with a lower volatility than Early-DI or PFI engines, but generally require a relatively higher reactivity.

ACKNOWLEDGMENTS

The authors would like to express their appreciation towards the SwRI Internal Research and Development Program advisors for supporting the work presented in this paper. We would also like to thank Cummins and Delphi Automotive Systems for providing hardware support for this project.

REFERENCES

1. Cheng, W. K., D. Hamrin, J. B. Heywood, S. Hochgreb, K. Min, and M. Norris (1993) "An Overview of Hydrocarbon Emissions Mechanisms in Spark-Ignition Engines" SAE Paper No. 932708.
2. Christensen, M. A., Hultqvist, and B. Johansson (1999) "Demonstrating the Multi-Fuel Capability on a Homogeneous Charge Engine with Variable Compression Ratio," SAE Paper No. 1999-01-3679.
3. Furutani M. Y. Ohta, M. Kono, and M. Hasegawa (1998) "An Ultra-Lean Premixed Compression Ignition Engine Concept and its Characteristics" Proceedings of the Fourth International Symposium COMODIA 98, pp.173-177.
4. Gale, N., D. W. Dickey, T. W. Ryan III, and R. H. Stanglmaier (2001) "HCCI Combustion Control in a Multi-Cylinder Engine through Dual Fuel Operation" JSAE Paper No. 20015341, presented at the Japanese Society of Automotive Engineers' Spring Conference, May of 2001.
5. Gray, W. A. III, and T. W. Ryan III (1997) "Homogeneous Charge Compression Ignition (HCCI) Combustion of Diesel Fuel," SAE Paper No. 971676.
6. Law, D., D. Kemp, J. Allen, G. Kirkpatrick, and T. Copland (2001) "Controlled Combustion in an IC-Engine with a Fully Variable Valve Train" SAE Paper No. 2001-01-0251.
7. Morimoto, S. S., Y. Kawabata, T. Sakurai, and T. Amando, (2001) "Operating Characteristics of a Natural Gas-Fired Homogeneous Charge Compression

Ignition Engine (Performance Improvement Using EGR)" SAE Paper No. 2001-01-1034.

8. Nakagome K., N. Shimazaki, K. Niimura, and S. Kobayashi, (1997) "Combustion and Emission Characteristics of Premixed Lean Diesel Combustion" SAE Paper 970898.
9. Olsson, J.-O., P. Tunestal, and B. Johansson (2001) "Closed-Loop Control of an HCCCI Engine" SAE Paper No. 2001-01-1031.
10. Stanglmaier, R. H., M. J. Hall, and R. D. Matthews (1997) "In-Cylinder Fuel Transport During the Cranking Cycles in a Port-Injected 4-Valve Engine" SAE Paper No. 970043.
11. Stanglmaier R. H. and C. E. Roberts (1999) "Homogeneous Charge Compression Ignition (HCCI): Benefits, Compromises, and Future Engine Applications", SAE Paper No. 1999-01-3682.
12. Stanglmaier R. H., J. Li, and R. D. Matthews (1999) "The Effect of In-Cylinder Wall Wetting Location on the HC Emissions from SI Engines" SAE Paper No. 1999-01-0502.
13. Stanglmaier, R. H., T. W. Ryan III, and J.S. Souder (2001) "HCCI Operation of a Dual-Fuel Natural Gas Engine for Improved Fuel Efficiency and Ultra-Low NO_x Emissions at Low to Moderate Engine Loads" SAE paper No. 2001-01-1897.
14. Takeda,Y., K. Nakagome, and K. Niimura (1996) "Emissions Characteristics of Premixed Lean Diesal Combustion with Extremely Early Staged Injection" SAE Paper No. 961163.

DEFINITIONS, ACRONYMS, AND ABBREVIATIONS

HCCI	Homogeneous Charge Compression Ignition. Used in this context to encompass all pre-mixed charge compression ignition combustion modes.
PCCI	Pre-mixed Charge Compression Ignition
CAI	Controlled Auto Ignition
PFI	Port Fuel Injection
Early-DI	Direct injection early in the engine cycle (intake stroke or early compression)
Late-DI	Direct injection late in the engine cycle (near TDC or shortly before combustion)
Reactivity	This term is used to indicate in general terms the propensity of a fuel for low-temperature auto-ignition. A fuel with lower reactivity requires a higher temperature for auto-ignition.

A New Generation of Engine Combustion Processes for the Future?
© P. Duret (Editor) and Editions Technip, Paris, 2001, pp. 77 - 84
27 rue Ginoux, 75015 Paris

PARAMETRIC STUDY OF GASOLINE HCCI WITH VARIOUS COMPRESSION RATIOS, INTAKE PRESSURES AND TEMPERATURES

Tomonori Urushihara, Koji Hiraya, Akihiko Kakuhou, Teruyuki Itoh
Nissan Motor Co., Ltd.

ABSTRACT

Single-cylinder test engine experiments were carried out to clarify the conditions under which stable and moderate HCCI (Homogeneous Charge Compression Ignition) combustion can be achieved. The experiments were performed under various compression ratios, intake pressures and intake air temperatures using standard gasoline as the test fuel. A region of stable and moderate combustion that satisfied both the limitation of CPi (Coefficient of variation of IMEP) and smoothness of combustion was found to have a wedge-like shape on the intake air temperature and air-fuel ratio plane. The region was confined by the CPi limitation line and the knocking line. The results showed that ignition timing was strongly affected by the intake air temperature and that the combustion duration, which needs to be considerably long for moderate combustion, was affected mainly by the air-fuel ratio. Consequently, the maximum fuel amount that can be input into the cylinder, and therefore, the maximum IMEP (Indicated Mean Effective Pressure) were determined according to the combination of the compression ratio and the engine speed. At high engine speeds where a short ignition delay is required, a higher intake air temperature was needed. This limited the maximum IMEP to a considerably low level in comparison with that of the low engine speed region. Supercharging was found to be one of the most promising ways of recovering IMEP in this area.

INTRODUCTION

Homegeneous Charge Compression Ignition (HCCI) is well known as a new combustion process for obtaining ultra-low NOx emissions with an exceptionally lean air-fuel ratio. Although HCCI has the potential to provide good engine performance, combustion instability at low engine loads and rapid heat release accompanied by unacceptable noise at high engine loads have been the basic problems in HCCI combustion. In many previous papers, these topics have been discussed both for gasoline [1-2] and diesel fuel [3-4].

In some previous studies, researchers tried to specify the conditions causing combustion instability or rapid combustion with an excessive heat release ratio. Thring [5] described the knocking region and misfire region on the EGR ratio–equivalence ratio plane. A satisfactory operating region free of both knocking and misfiring was reported to be located in between these two areas. Oakley et al. [6] successfully obtained a CAI (Controlled Auto-ignition) region that was free of both knocking and misfiring on the EGR ratio–excess air ratio plane.

Although these two-dimensional explanations are visually understandable, there seems to be some difficulty in explaining phenomena by using the EGR ratio as one of the two axes. EGR reduces the oxygen concentration of the in-cylinder mixture and also has an effect on suppressing the occurrence of auto-ignition. Besides these effects, EGR increases the heat capacity, which decreases the adiabatic flame temperature and suppresses the heat release ratio. Because of these combined effects of EGR, selecting some parameter other than the EGR ratio for the second axis might facilitate a simpler explanation.

Expanding the HCCI operating region both in terms of engine speed and engine torque is very important when considering the use of HCCI in production vehicles. Enlarging the HCCI operating region towards higher IMEP is thought to be especially important for attaining the maximum fuel economy benefit from HCCI combustion by utilizing HCCI as much as possible throughout driving cycles.

Aoyama et al. [7] obtained relatively high IMEP of 0.6 MPa among the gasoline HCCI combustion studies reported previously by using a high compression ratio (17.4:1) in a single-cylinder engine at a low engine speed of 1000 rpm. The engine speed, however, was not varied in their research. Ishibashi et al. [8] succeeded in running their two-stroke "ATAC" engine on gasoline over a wide engine speed range from 2000 to over 8000 rpm. Their maximum BMEP was about 0.35 MPa, which was lower than the load of the preceding 4-stroke engine throughout the entire engine speed range. Although there are obvious differences in the operatiing range between HCCI combustion achieved in a high compression ratio 4-stroke engine and that obtained in a 2-stroke "ATAC" engine, little attempt has been made to explain such differences comprehensively.

In this study, single-cylinder test engine experiments were conducted to clarify the conditions under which stable and moderate HCCI combustion can be accomplished. Various compression ratios, intake pressures and intake air temperatures were tested using standard gasoline as the test fuel. The data were plotted on a two-dimensional plane using the intake air temperature as one axis and the air-fuel ratio as the other axis. A region of stable and moderate combustion limited by the allowable CPi (Coefficient of variation of IMEP) line and the knocking line was found to have a wedge-like shape. The maximum fuel amount that can be input into the cylinder, and therefore, the maximum load, can be derived from the tip of the wedge-shaped region. Some possibilities for improving the maximum IMEP with respect to this point are also discussed in this paper.

TEST ENGINE

The test engine was a four-valve-per-cylinder unit having a pentroof combustion chamber and other major specifications as shown in Table 1. The compression ratio was varied from 12:1 to 18:1 in order to recreate the typical two-stroke engine HCCI combustion described in reference [4] and the typical four-stroke engine HCCI combustion discussed in reference [3]. The intake air temperature was also varied from ambient temperature to 250 deg. C by means of an electrical heater inserted in the intake pipe. Gasoline was injected into the intake port from a fuel injector mounted downstream of the intake manifold. A fuel injection timing of top dead center of the compression stroke was selected so as to maximize fuel vaporization in the intake port.

Bore, Stroke	86 mm, 86 mm
Displacement Vol.	500 cc
Fuel Supply	Port Fuel Injection
Compression Ratio	12, 13.5, 15, 16.5, 18
Engine Speed	1200, 2400 rpm
Intake Pressure	101, 152 kPa

Table 1 Test Engine Specifications

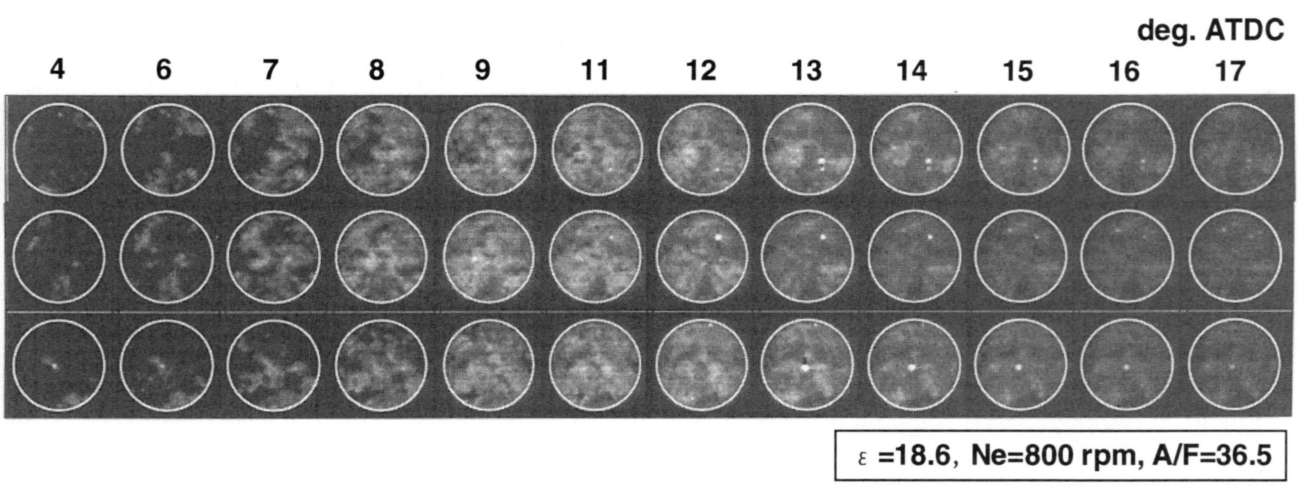

Fig. 1 High-Speed Photographs of HCCI Combustion

RESULTS AND DISCUSSION

HIGH-SPEED PHOTOGRAPHS OF HCCI COMBUSTION -

High-speed photographs of HCCI combustion taken in three consecutive cycles are shown in Fig. 1. Combustion was stable and moderate under conditions of 800 rpm and an A/F ratio of 36.5:1 with a compression ratio of 18.6:1 where the photographs were taken. There is a certain inhomogeneity in the flame distribution especially at the beginning of combustion (4-8 deg. ATDC). After that period, however, the flame distribution shows more homogeneity from 9-17 deg. ATDC. This indicates that combustion at each local point was not instantaneous but proceeded gradually. The overall combustion duration is determined by both the local combustion period and combustion phasing across the combustion chamber. To achieve moderate HCCI combustion, it is important to slow down the local combustion speed. In order to achieve stable combustion, which is the other concern with HCCI, ignition needs to be controlled so that it occurs near TDC.

STABLE AND MODERATE COMBUSTION REGION -

To clarify the basic characteristics of HCCI, both the intake air temperature and the air-fuel ratio were varied independently under conditions of a compression ratio of 15:1, an engine speed of 1200 rpm and natural aspiration.

Figure 2 shows the moderate and stable combustion region on the intake air temperature--air-fuel ratio plane. The plane is separated into two regions. One is a region in which combustion proceeds steadily, located in the area of a high intake air temperature and a rich air-fuel ratio. The other is a region in which combustion is not stable or does not even occur. The former region is again divided into a knocking region and a moderate combustion region, which accounts for roughly the upper half of the region of steadily proceeding combustion. Consequently, the region of stable and moderate combustion, which is limited by the allowable CPi line and the knocking line, is seen to have a wedge-like shape.

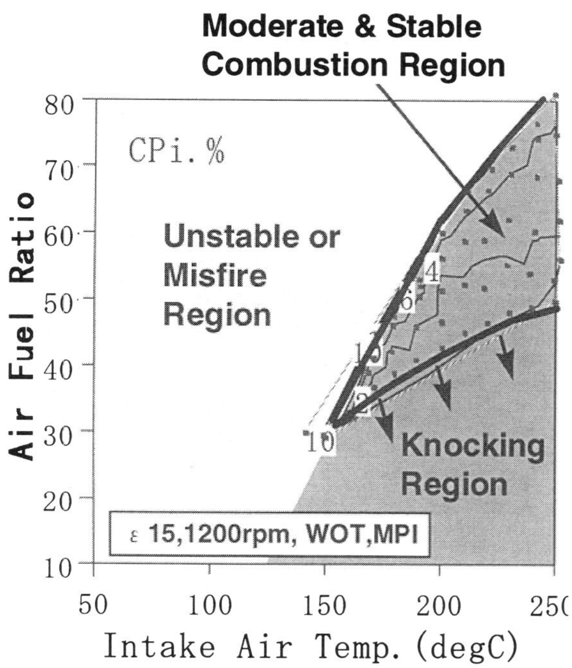

Fig. 2 Stable and Moderate Combustion Region

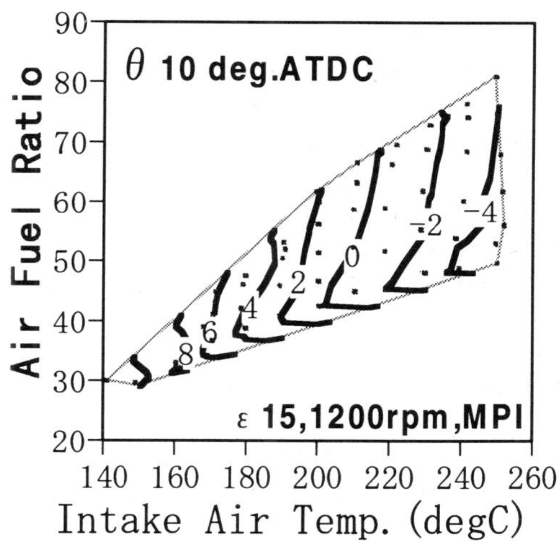

(a) Contours of 10% mass burnt

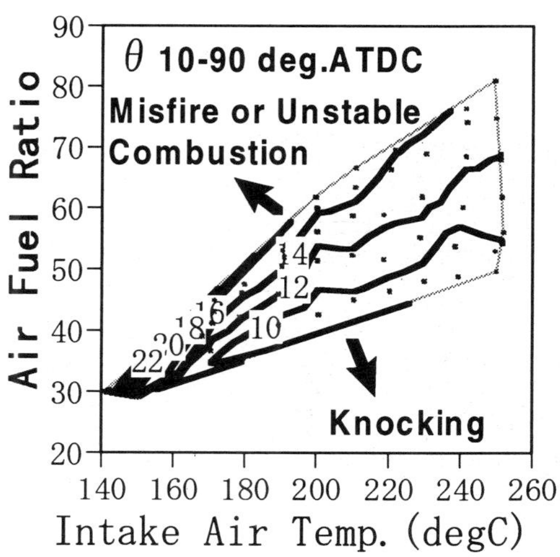

(b) Contours of combustion duration from 10% to 90% mass burnt

Fig. 3 Effect of Intake Air Temperature and Air-Fuel Ratio on Auto-ignition Timing and on Combustion Duration

IGNITION TIMING AND COMBUSTION DURATION -

Ignition timing - Figure 3 shows contours of the 10% mass burnt crank angle (Fig. 3(a)) and main combustion duration (Fig. 3(b)), which is defined as the period from 10-90% mass burnt. In Fig. 3(a), it is seen that the contours of ignition timing are almost vertical. This means that the time when auto-ignition occurs can be adjusted effectively by the intake air temperature.

Combustion duration - On the other hand, the combustion duration contours in Fig. 3(b) show a relatively horizontal inclination. This means that the combustion duration is mainly affected by the air-fuel ratio.
This phenomenon is explained by the mechanism shown in Fig. 4. The chemical reaction rate is strongly dependent on temperature. During combustion, local heat release raises the local temperature. The higher local temperature further accelerates local heat release. Thus, there is a positive feedback loop in HCCI combustion. The feedback coefficient, i.e., the ratio between the temperature rise and the released heat is inversely proportional to the heat capacity of the mixture with respect to the unit mass of fuel. Accordingly, an increase in heat capacity, in other words, an increase in the air-fuel ratio, reduces the feedback coefficient. Thus, increasing the air-fuel ratio is effective in achieving moderate HCCI combustion.

EXHAUST EMISSIONS AND THERMAL EFFICIENCY -

Exhaust emissions - Unburned HC and CO concentrations in the exhaust gas are shown in Fig. 5(a) and Fig. 5(b), respectively. The HC concentration decreases with an increase in the intake air temperature, and the CO concentration increases with an increase in the air-fuel ratio. It is seen that CO, which is an intermediate product of combustion, is not completely oxidized when the adiabatic flame temperature is low.

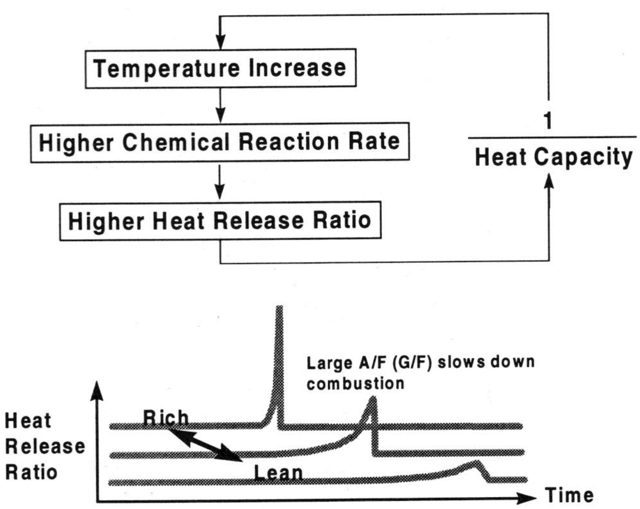

Fig. 4 Feedback Mechanism in HCCI Combustion

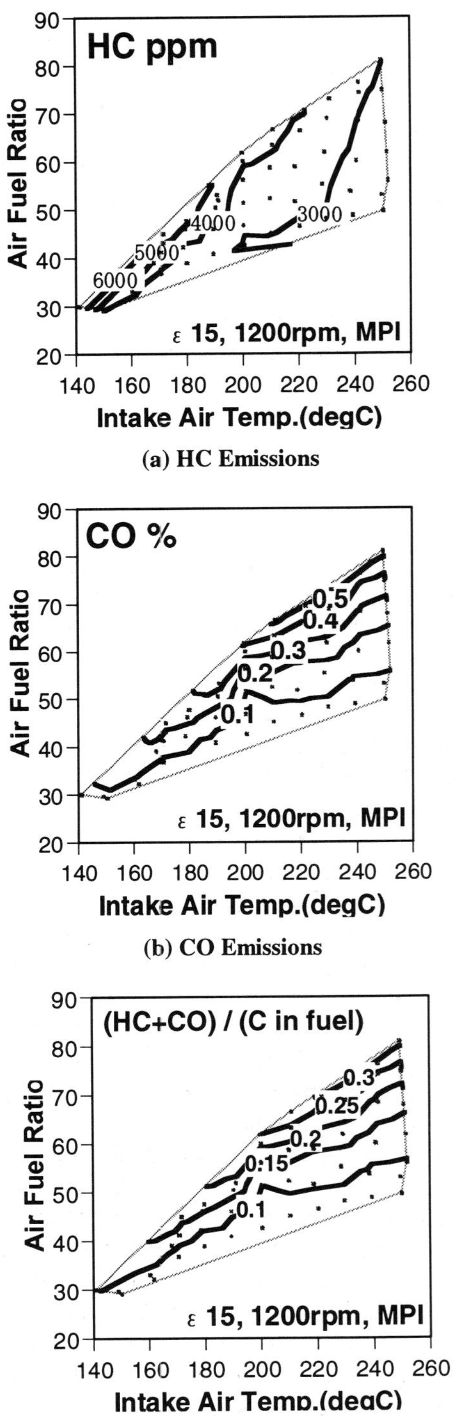

(a) HC Emissions

(b) CO Emissions

(c) Incomplete Combustion Ratio

Fig. 5 HC and CO Emissions

Figure 5(c) shows the ratio of the carbon emitted as HC and CO in the exhaust gas to the carbon content in the fuel. In the lean air-fuel mixture region, more than 30% of the carbon in the fuel was emitted at a maximum without being oxidized completely even in the stable combustion region. This affects thermal efficiency.

Thermal efficiency - The ratio of heat loss to total fuel energy is shown in Fig. 6. The heat loss ratio increases with an increase in the intake air temperature and with a decrease of the air-fuel ratio. The reason why the heat loss ratio increases with a decrease of the air-fuel ratio is thought to be that the peak time of heat release advances with an enriched air-fuel ratio.

Finally, the derived thermal efficiency is shown in Fig. 7. The best thermal efficiency is obtained at the tip of the wedge where the largest combustion efficiency and the smallest heat loss ratio are obtained.

MAXIMUM LOAD -

Maximum load point in the stable and moderate combustion region - Since the intake air temperature determines charging efficiency, and the combination of charging efficiency and the air-fuel ratio determines the fuel quantity injected into the cylinder, the fuel mass is determined uniquely at each point on the intake air temperature and air-fuel ratio plane. This means that IMEP at each point on the plane is also approximately determined uniquely in the same way. The closer to the origin the point is, the higher IMEP becomes.

Figure 8 shows IMEP contours on the intake air temperature and air-fuel ratio plane. Because of the reason mentioned above, the maximum IMEP is obtained at the tip of the wedge region. To improve the maximum IMEP, i.e., to move the tip toward the origin, there are obviously two ways. One is to expand the knocking limit on the rich air-fuel ratio side by increasing the heat capacity of the mixture with respect to the unit mass of fuel. Supercharging the inert gas is clearly one of the ways to accomplish that. The other is to expand the misfire limit on the lean air-fuel ratio side by improving the ignitability of the mixture. One obvious way to accomplish that is to utilize a low octane number fuel. Besides changing the fuel, increasing the compression ratio is another possible way of improving mixture ignitability.

As for the minimum IMEP, there is no limitation as long as the intake air temperature can be varied

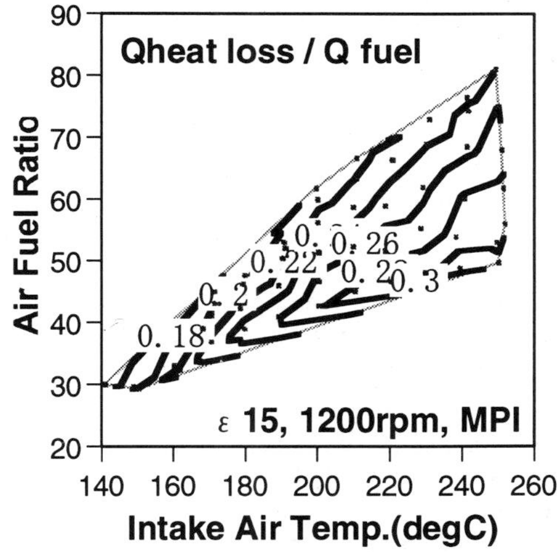

Fig. 6 Ratio of Heat Loss to Supplied Fuel

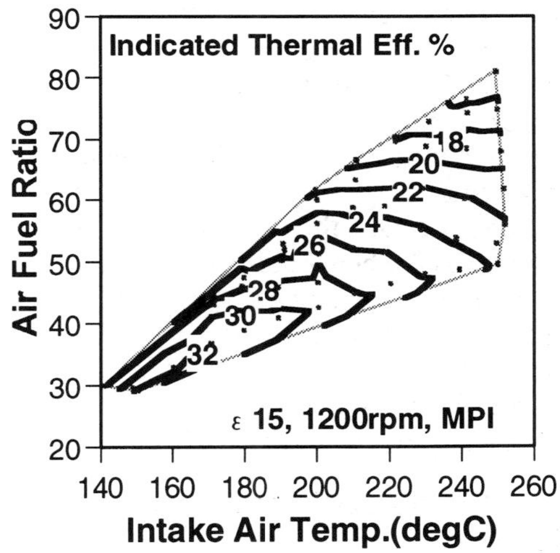

Fig. 7 Indicated Thermal Efficiency

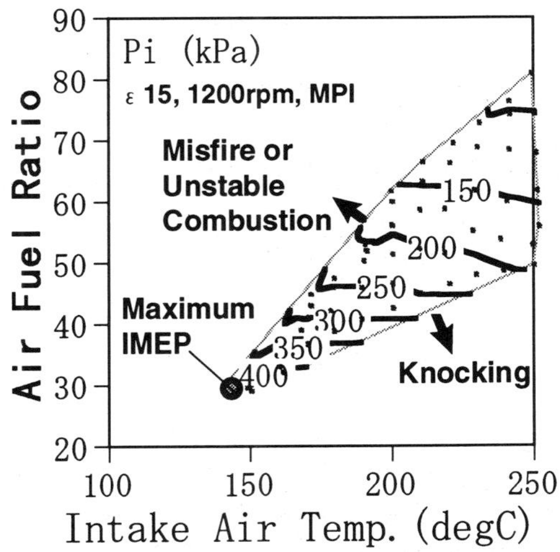

Fig. 8 Indicated Mean Effective Pressure

arbitrarily. The disadvantage of using a high intake air temperature to obtain a low engine load is an increase in heat loss as shown in Fig. 6.

Improvement of maximum load - As mentioned above, increasing the compression ratio is one way to improve ignitability without intake air heating, which causes charging efficiency to drop. Figure 9 shows how the compression ratio affects the wedge-shaped region, i.e., the stable and moderate combustion region. The location of the wedge-shaped region shifts to the left as the compression ratio is increased. The reason for this is that a certain mixture temperature near compression TDC is obtained at a lower intake air temperature by increasing the compression ratio. As a result of the shift in the location of the wedge-shaped region, the maximum IMEP is improved from 450 to 540 kPa by increasing the compression ratio from 15:1 to 18:1.

The other way to improve the maximum IMEP is supercharging as mentioned above. Figure 10 shows the location of the wedge-shaped region corresponding to each compression ratio on the plane under an intake air pressure of 152 kPa. In the case of a compression ratio of 15:1, the maximum IMEP is improved from 450 to 740 kPa by increasing the intake pressure from 101 to 152 kPa (abs.). Increasing the compression ratio beyond 15:1 does not improve the maximum IMEP in contrast to the result seen for natural aspiration. The reason is that the intake air temperature at the wedge tip falls below atmospheric temperature under a combination of supercharging and a high compression ratio such as 18:1. In this study, such a low temperature was not utilized.

Effect of engine speed on the maximum engine load - Figure 11 shows the location of the wedge-shaped region corresponding to each compression ratio on the plane when the engine speed was raised to 2400 rpm. A higher engine speed requires a shorter ignition delay to obtain the proper ignition timing for stable combustion. That means a higher intake air temperature is required at higher engine speeds. Therefore, the CPi limit lines shifted toward a higher temperature than in the case of 1200 rpm, while the knocking lines hardly moved. Consequently, the wedge-shaped regions shifted in the direction of a higher intake air temperature and a leaner air-fuel ratio. The maximum IMEP decreased at every compression ratio.

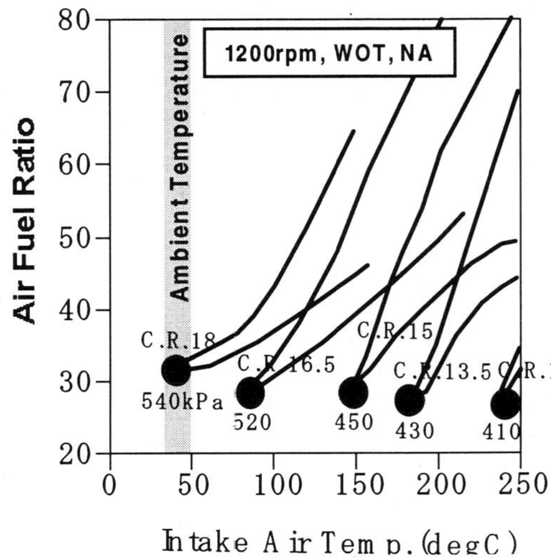

Fig. 9 Effect of Compression Ratio on the Wedge-shaped Region and on the Maximum IMEP

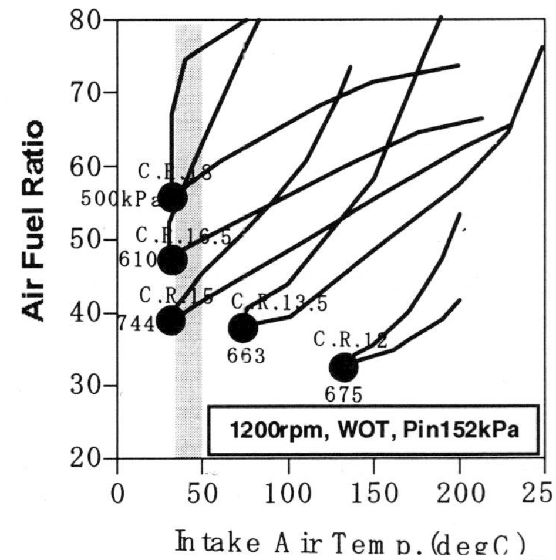

Fig. 10 Effect of Supercharging on the Wedge-shaped Region and on the Maximum IMEP

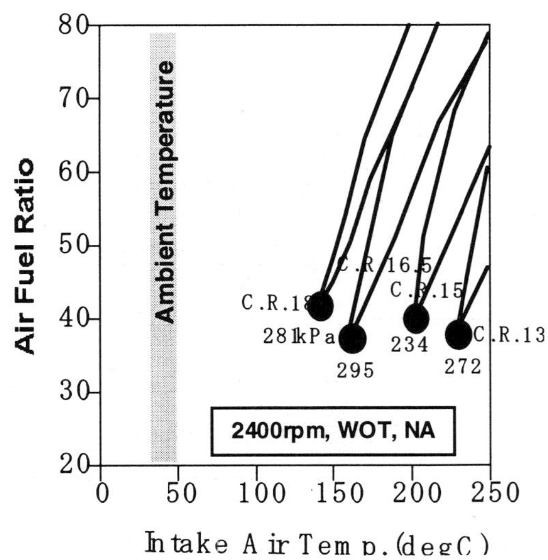

Fig. 11 Effect of Engine Speed on the Wedge-shaped Region and on the Maximum IMEP

ENGINE OPERATING RANGE -

Figure 12 plots the maximum IMEP obtained by using a compression ratio of 18:1 without any charge heating and the maximum IMEP obtained by using a compression ratio of 12:1 with charge heating. A compression ratio of 18:1 without charge heating was selected as being typical of four-stroke engine HCCI discussed, for example, in reference [7]. And a compression ratio of 12:1 with charge heating was selected to represent two-stroke engine HCCI described, for example, in reference [8]. With only a high compression ratio, the engine speed did not go above 1200 rpm. In contrast, with intake air heating, the operating range expanded toward a higher engine speed even at a low compression ratio such as 12:1. However, the maximum IMEP was limited especially at higher engine speeds by the higher intake air temperature that was required, as was explained in the previous section.

Results obtained with internal EGR, which was utilized as the heat source in place of intake air heating, are also plotted in Fig. 12. The maximum IMEP values attained were nearly identical to those obtained by intake air heating. The reason for this agreement needs further analysis in order to be understood.

CONCLUSIONS

Single-cylinder test engine experiments were carried out to clarify the conditions under which stable and moderate HCCI combustion is achieved under various compression ratios, intake pressures and temperatures using gasoline and port fuel injection.

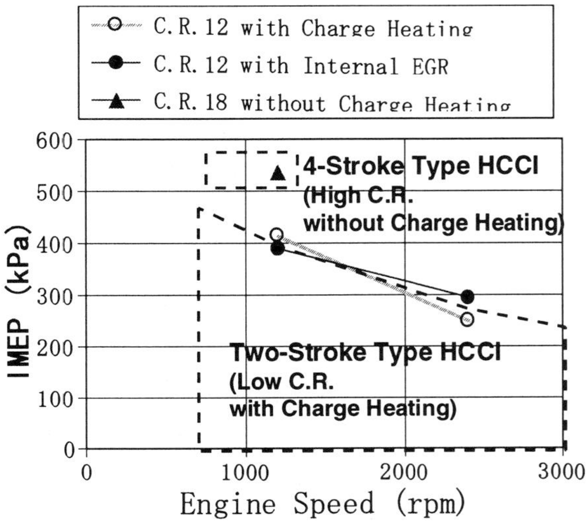

Fig. 12 Examples of Engine Operating Point achieved with HCCI

The results obtained are summarized below.

(1) The region of stable and moderate combustion was found to have a wedge-like shape on the intake air temperature and air-fuel ratio plane.

(2) The ignition timing is strongly affected by the mixture temperature, and the combustion duration is affected mainly by the air-fuel ratio in the region.

(3) The maximum IMEP is obtained at the tip of the wedge-shaped region. The value of the maximum IMEP is determined by the combination of the compression ratio and engine speed.

(4) Either supercharging or raising the compression ratio was found to be essentially effective in improving the maximum IMEP.

ACKNOWLEDGEMENTS

The authors wish to express their thanks to Mr. Tatsumi Iga and Mr. Takashi Naka for their cooperation with the experiments.

REFERENCES

1. P. M. Najt and D. E. Foster, "Compression-Ignited Homogeneous Charge Combustion", SAE Paper 830264, 1983.
2. J. Lavy, J. Dabadie, et al., "Innovative Ultra–low NOx Controlled Auto-Ignition Combustion Process for Gasoline Engines: the 4-SPACE Project", SAE Paper 2000-01-1837, 2000.
3. H. Akagawa, T. Miyamoto, et al., "Approaches to Solve Problems of the Premixed Lean Diesel Combustion", SAE Paper 1999-01-0183, 1999.
4. N. Shimazaki, H. Akagawa and K. Tsujimura, "An Experimental Study of Premixed Lean Diesel Combustion", SAE Paper 1999-01-0181,1999.
5. R. H. Thring, "Homogeneous-Charge Compression-Ignition (HCCI) Engines", SAE Paper 892068, 1989.
6. A. Oakley, H. Zhao and N. Ladommatos, "Experimental Studies on Controlled Auto-ignition(CAI) Combustion of Gasoline in a 4-Stroke Engine", SAE Paper 2001-01-1030, 2001.
7. T. Aoyama, et al., "An Experimental Study on Premixed-Charge Compression Ignition Gasoline Engine", SAE Paper 960081, 1996.
8. Y. Ishibashi and M. Asai (1999), "Improving the Exhaust Emissions of Two-Stroke Engines by Applying the Activated Radical Combustion", SAE Paper 960742, 1996.

A New Generation of Engine Combustion Processes for the Future?
© P. Duret (Editor) and Editions Technip, Paris, 2001, pp. 85 - 100
27 rue Ginoux, 75015 Paris

ADVANCED COMBUSTION USING A LOTUS ACTIVE VALVE TRAIN

INTERNAL EXHAUST GAS RECIRCULATION PROMOTED AUTO-IGNITION

Jeff Allen, Don Law
Lotus Engineering

ABSTRACT

Auto-ignition is a phenomenon, which holds distinct promise in 4-stroke engines. Its primary benefits are those of reduced fuel consumption and considerably reduced pollutants. Previously, workers in this field have been forced to depend on pre-heating the intake charge (40-200°C) in order to make auto-ignition work. We have shown conclusively that pre-heating the intake charge is not required if internal exhaust gas recirculation (EGR) is used instead. To accomplish this advanced valve strategies have been employed using the Lotus active valve train (AVT), whereby, we see auto-ignition phenomena. Examples of such valve strategies will be discussed together with a presentation of the speed and load ranges over which auto-ignition can be utilised and how the advanced valve strategies adapt to specific speed-load ranges. Emissions and fuel economy maps will also be presented together with images caught using a single cylinder optical engine showing the nature of the combustion in-cylinder and so differentiating it from conventional spark ignition.

INTRODUCTION

For some time now Lotus has been one of the pioneers of fully variable valve trains (FVVT). Much of this work has centered on the Lotus Active Valve Train (AVT) [1-3], which is widely recognised as an advanced combustion and cam profiling research and valve operating strategy development tool. Recently results have been reported with other electro-mechanical systems [4] and electro-hydraulic systems [5] although not without limitations in the flexibility of their operation.

It is logical to assume, that before long, modular production electrically controlled FVVT's will become available and eventually commonplace. The Lotus AVT system has allowed us to probe such camless horizons with innovative combustion methodologies.

This paper presents results from a combustion process that we refer to as a form of advanced combustion (AC). This process is called controlled auto-ignition (CAI). Controlled auto-ignition is well recognised in the 2-stroke world [6-16]. Reports in the 4-stroke world [17-18] were quite sparse and again found to be limited by either the limited flexibility of their valve control events, using fixed profile cams or a dominant need for the intake air: fuel mixture to be pre-heated (between 50-200°C) in order to achieve CAI. IFP [18] first reported CAI without pre-heating of the charge and using internal EGR.

This work was followed by Lotus [19, 20] where the first demonstration of internal EGR in CAI was reported using a *true* FVVT. The work of IFP and Lotus follows that of others [21-23] who made valuable experimental contributions to this form of combustion phenomena. A valuable attempt to map the operating region of CAI has recently been reported although again using charge pre-heating [24]. Our work presented here presents the first operating maps for gasoline

CAI using a FVVT. The CAI combustion process operates without spark by utilising internal EGR achieved through the total valve event control of the AVT. Initial data shows that the CAI process reduces NOx and CO emissions compared to conventional spark ignition (SI) engines in specific speed load regimes, principally at part-load conditions. Once FVVT's become widely accepted, then we envisage great synergy between AC (in the form of CAI) and SI.

EXPERIMENTAL ACTIVE VALVE TRAIN (AVT)

AVT is an electrically controlled, hydraulically actuated engine valve system. A cutaway section of a research grade system shown in Figure 1. The four engine valves are hydraulically actuated by pistons, driven via fluid flow, precisely metered using high frequency switching Moog valves. The engine valve stem is connected directly to the hydraulic force authority, which gives the greatest degree of control and flexibility. Precise dynamic valve control is accomplished using linear displacement transducers (LDT's) connected to a shaft extended from the top of the engine poppet valve.

Figure 1: The research grade AVT system.

Control software is constantly aware of valve position, relative to any piston via closed-loop control with a crankshaft encoder. Desired valve events (or profiles) are entered into the control software by the user and the LDT control algorithm then operates the valves until the *desired* profile versus *actual* controlled valve motion match. Differences between desired versus actual can then be eliminated using specific valve gain controllers, thus producing extremely accurate valve profiles. As we move towards the next generation of engines, with increasing pressure from legislation and customers, the use of variable valve systems will increase.

Figure 2: Non-optical single cylinder research engine fitted with the active valve train (AVT). The engine capacity is 0.45L and limited to a maximum operating speed of 5000rpm.

Lotus has therefore taken the well-proven research AVT system and is now developing it into a mainstream production version. A number of basic changes have been made to the system architecture to make the system production viable in high volumes, but the basic concept of a true FVVT has been maintained, as the authors believe this is the correct solution for compliance with future legislative and customer demands as well as being the key enabler for more conventional and new advanced combustion strategies. The production AVT system now incorporates a conventional valve spring which both holds the valves shut on engine shut down and eliminates the need for a double acting hydraulic cylinder thus reducing the cost of manufacture significantly. Specifications of the research and production system can be seen below in Table 1.

VALVE STRATEGIES

The AVT is a truly variable valve operating system, with flexibility in lift, duration and velocities. In Figure 3, the lift of each individual valve event can be controlled from 0-10mm or 15mm. Figure 4, indicates complete flexibility the system has in varying duration of any valve opening event, here the opening and closing points of each valve are precisely controlled which means phasing of valve events may also be altered, shown in Figure 5. Indicated in Figure 6, opening and closing velocities of each valve are also controllable giving this system the ability to change valve profiles from symmetrical polynomial, asymmetric or even square wave profiles. This function of velocity control using the displacement transducer for real time feed back of position also allows for a controlled 'soft touch down' feature on all valves. Indeed, the soft touch down event can be optimised at each speed and load condition to give the best closing velocity for NVH, durability and emission issues.

This true flexibility of valve events can now be exploited for the first time to extract the most efficient operating strategies from any internal combustion engine it is fitted to.

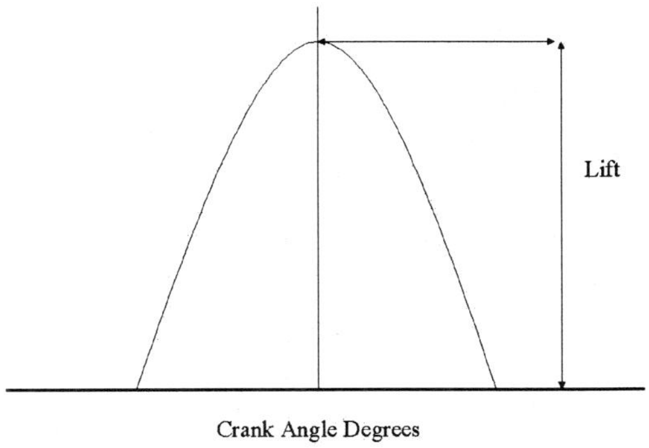

Figure 3

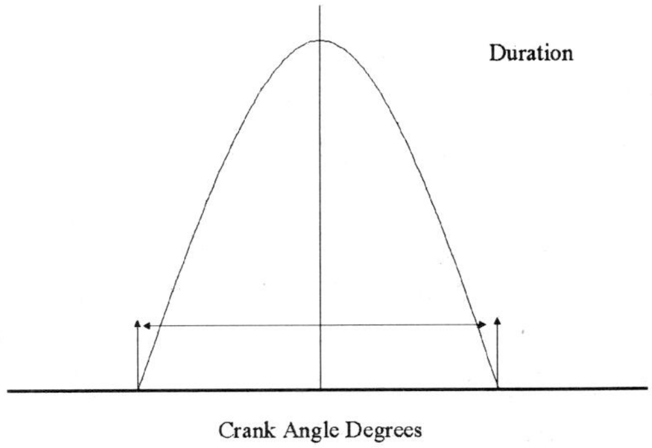

Figure 4

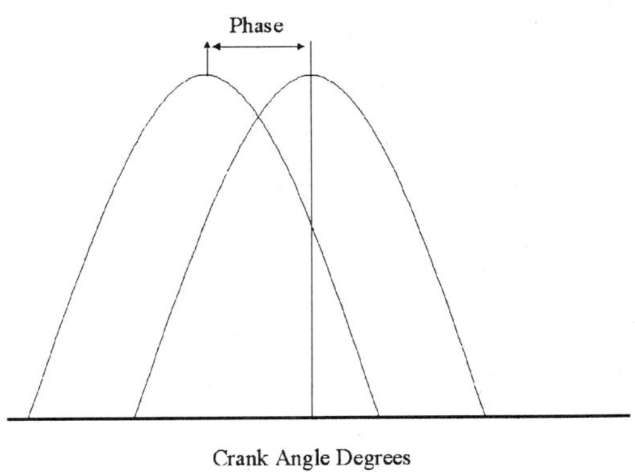

Figure 5

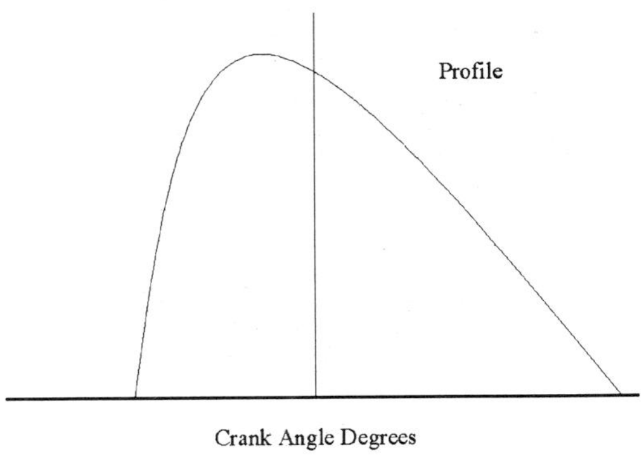

Figure 6

The next section describes three possible valve-operating modes that can achieve true throttleless operation and thus eliminate induction-pumping losses normally associated with SI engines. It is now possible to operate engine poppet valves to achieve true throttleless engine operation while still using clean, stable, homogeneous combustion modes. Shown in Figure 7, is a set of valve profiles used to achieve throttleless SI with a homogeneous charge. These are actual profiles recorded from the single cylinder engine seen in Figure 2 operating at 2000rpm. Profile A is the standard production representative polynomial profiles used to achieve a "full load" unthrottled condition. Profiles B through to F show progressively shorter duration profiles (at -10° intervals) affecting only intake valves, that gradually reduce the trapped fresh charge in the cylinder by advancing intake valve closing point and thus effectively and efficiently reduce the load of the engine. Profile F, has reduced the load of the engine to an effective "zero" load condition at 2000rpm and reducing the valve duration and lift further, profiles G, H, and I, now decrease the engine speed from 2000rpm to below 1000rpm resulting in profile I, giving a true throttleless idle condition for an SI engine.

An alternative throttleless valve strategy is shown graphically in Figure 8, and pictorially in Figure 9. This strategy achieves load control by varying the closing of the exhaust valves and the opening of the intake valves. Again this is a true throttleless operating strategy however this strategy, unlike that shown in Figure 7, displaces fresh charge with trapped exhaust gas residuals (EGR) and results in CAI combustion operation, discussed in greater detail later in the paper.

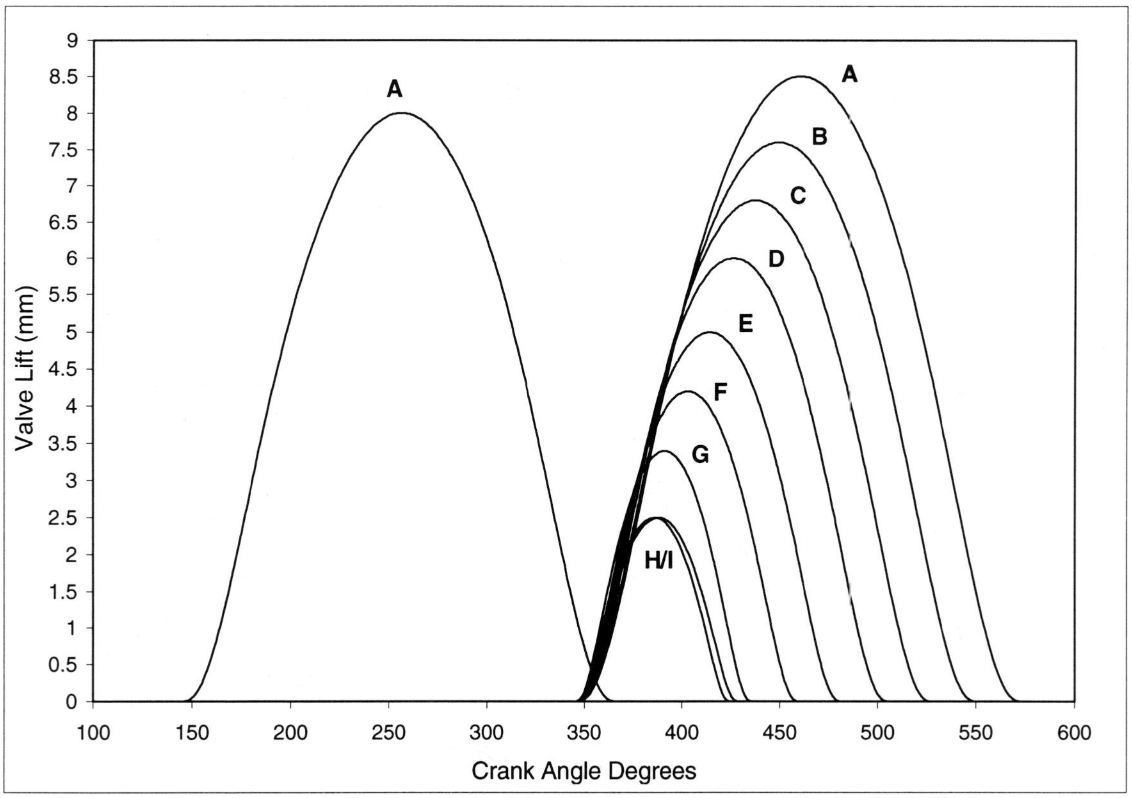

Figure 7

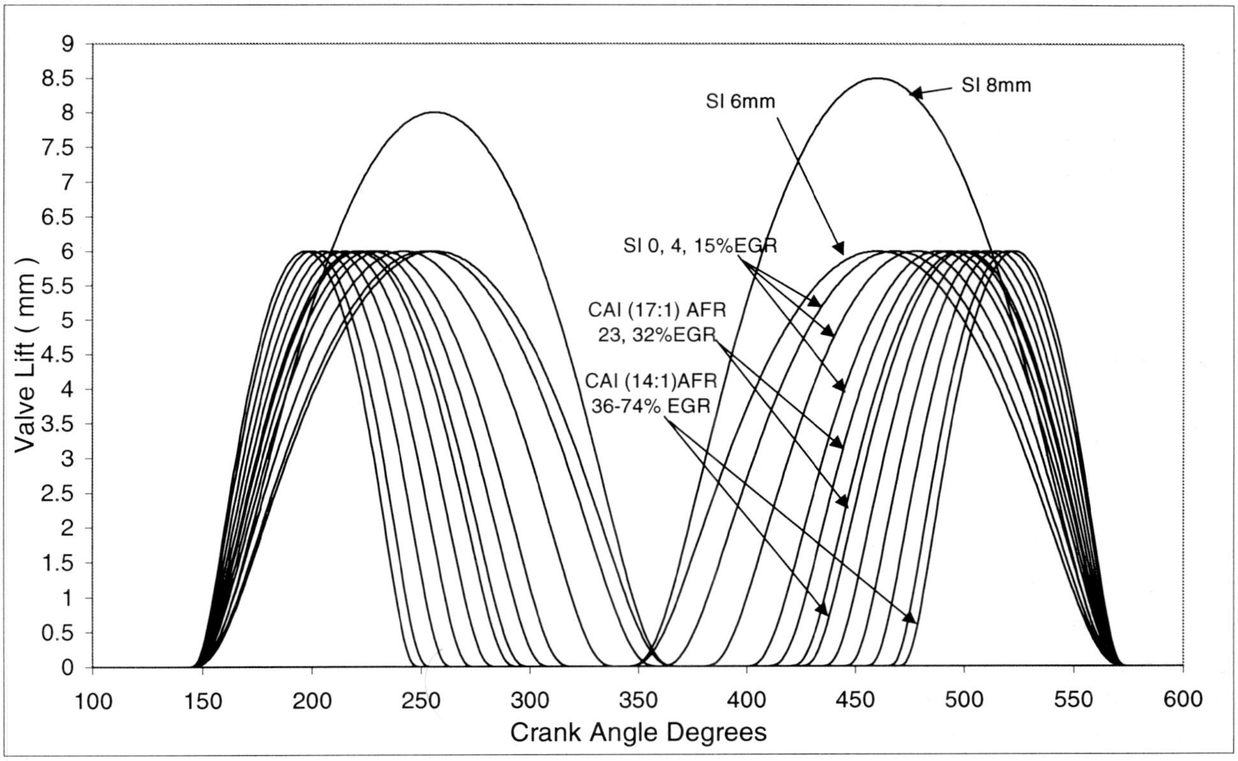

Figure 8

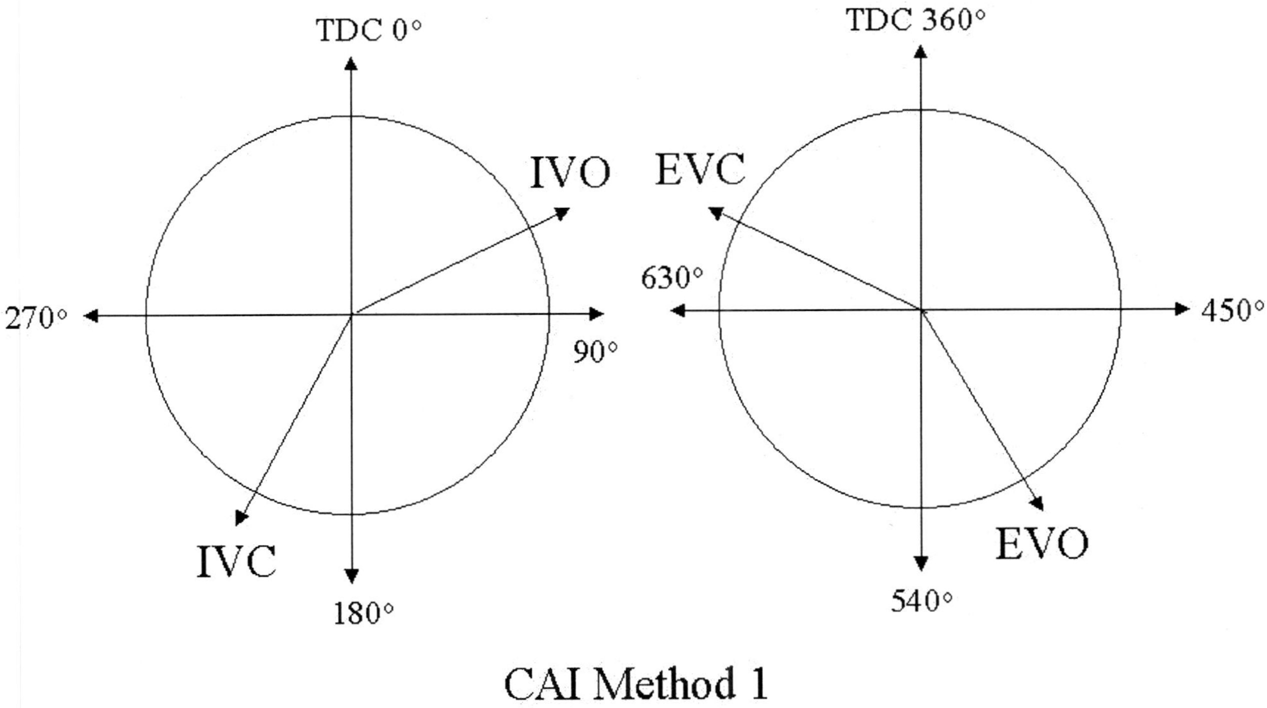

Figure 9

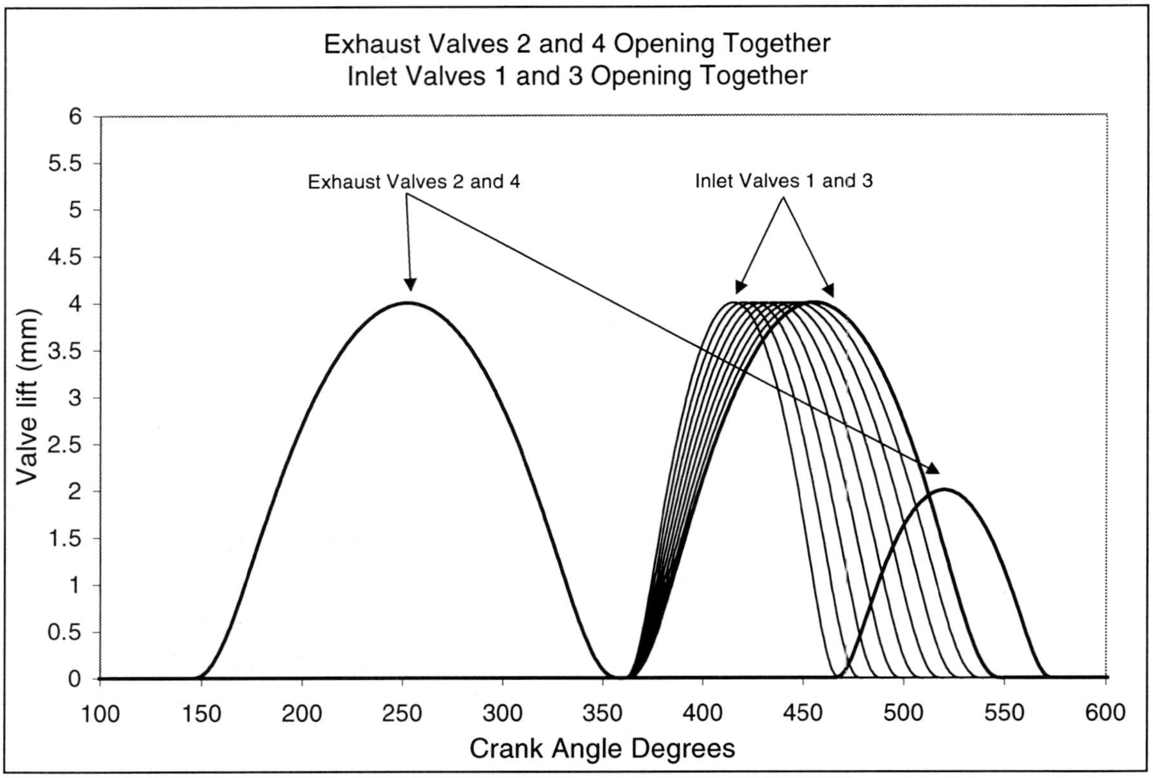

Figure 10

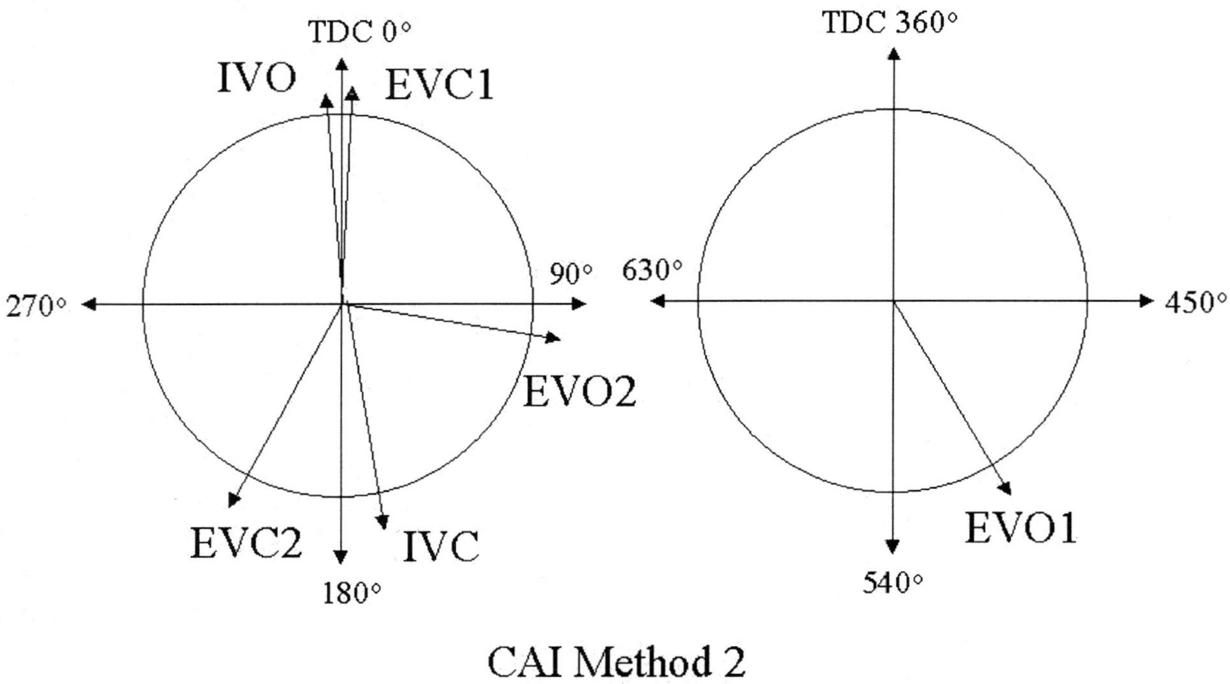

Figure 11

In Figure 8, the spread of valve profiles, again shows the transition from the full load SI (at 8mm lift) operation, through part load SI with increasing levels of EGR (SI 6mm lift with 0, 4 and 15% EGR), to the conditions with higher levels of EGR, which support CAI (23-74% EGR). In this method of throttleless operation, the cylinder is always filled with a gas mix and the load control is achieved through control of the proportion of EGR compared to fresh charge.
Figures 10 and 11 describe another throttleless valve operating strategy that also may be used to achieve CAI. Here the exhaust valve profiles are maintained at the standard 8mm lift for the first event, but are then both or individually re-opened for a second time part way through the induction stroke. Load control in this operating mode, is achieved through reducing the intake valve opening duration by advancing the intake valve closing point. This has the effect of reducing the amount of fresh charge admitted into the cylinder and therefore increasing the amount of EGR re-admitted into the cylinder during the second exhaust valve event, hence reducing engine load.

RESULTS AND SPEED LOAD MAPS

A single cylinder engine equipped with AVT has been operated using the valve strategies described in the preceding section and the results of those tests are shown in the following series of graphs. Results are all converted to break specific terms (e.g. brake mean effective pressure, bmep, brake specific fuel consumption, bsfc) to allow comparison with conventional engines. It must be borne in mind, however, that the engine is a single cylinder and therefore has a greater frictional force per cylinder than a multi cylinder engine.

The first set of graphs, Figures 12 to 15, are sets of data taken from the engine running with fixed standard production valve profiles (Table 3) and an intake throttle to control engine load. These data are baseline results that are used as a comparison throughout this paper, as they represent a current state-of-the-art multi point fuel injection homogeneous combustion Euro III engine.

Figures 16 to 19 show results from running the engine in a CAI strategy as illustrated in Figures 8 and 9. The results shown are restricted to the area of operation that represented stable stoichiometric CAI. While the engine was operated with the same valve strategy (early exhaust valve closing and late intake valve opening) outside of the areas shown in Figures 16 to 19 it was either run with lean CAI, spark assisted CAI or in pure SI when the EGR percentage dropped below 20%. Concentrating on the area of stoichiomtric CAI, Figures 16 to 19, we can see that the range is currently limited to speeds of 1000-4000rpm and loads from 0.5 bmep at higher speeds, and loads up to 4.5bar at low speeds. The benefits of CAI can be clearly seen in the comparison maps, Figures 20 to 23. Figure 20 shows the improvement in fuel economy achieved when operating the engine in CAI with reduced fuel consumption of up to 32% compared with baseline SI. Fuel economy reductions are achieved through out the operating range of CAI with greatest savings achieved load is reduced and speed increased. Fuel savings are, it is believed, mostly attributable to throttleless operation of CAI when using the AVT system. Figure 21, identifies the reductions in brake specific NO_X when operating in CAI and shows reductions across the complete operating range of this combustion strategy. Benefits of up to 97% at the lighter loads are very significant and are easily achievable with the FVVT. Figure 22, shows the improvements in total hydrocarbon (THC) emissions. As can be seen, the CAI combustion strategy gives reductions of up to 45%. This time however, the largest benefits are achieved at lower speeds and higher operating loads, with increases in THC emissions being evident at the lighter loads and higher speeds. Figure 23, shows reductions in break specific carbon monoxide (CO) with improvements across the operating range from 16 to 52%. The range of the CAI area is restricted by a set of in cylinder conditions that currently prevent CAI occurring at very low speeds and loads, at high engine speeds and at high engine loads. The limitation at low speeds and loads, is thought to be due to insufficient thermal levels in the cylinder due to the very small amount of fuel being burnt, this problem may be resolved on a multi cylinder with the use of features such as cylinder deactivation at idle that effectively increase the work done by the smaller number of cylinders still working. The limitation at high speeds and loads is due to uncontrolled detonation. This can be prevented to some extent by increasing the air/fuel ratio and operating in a lean CAI combustion strategy which allows higher loads to be achieved. At higher loads still, it is currently necessary to switch back to SI combustion.

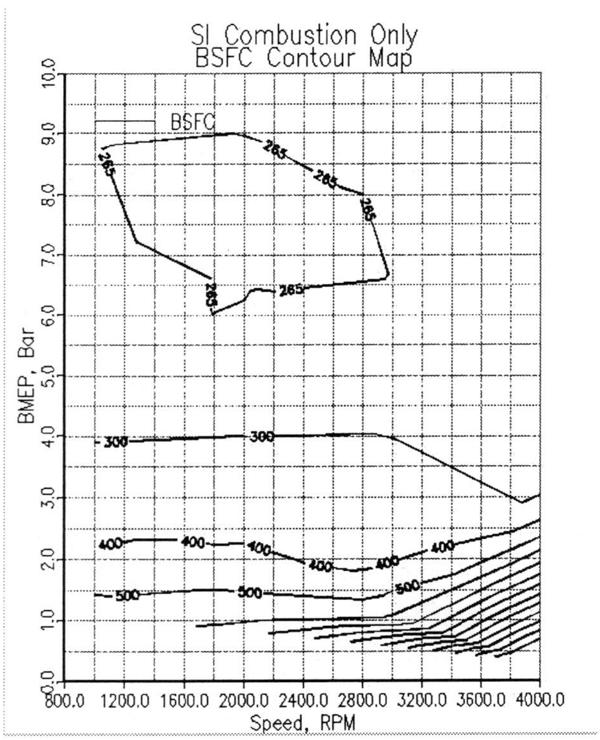

Figure 12

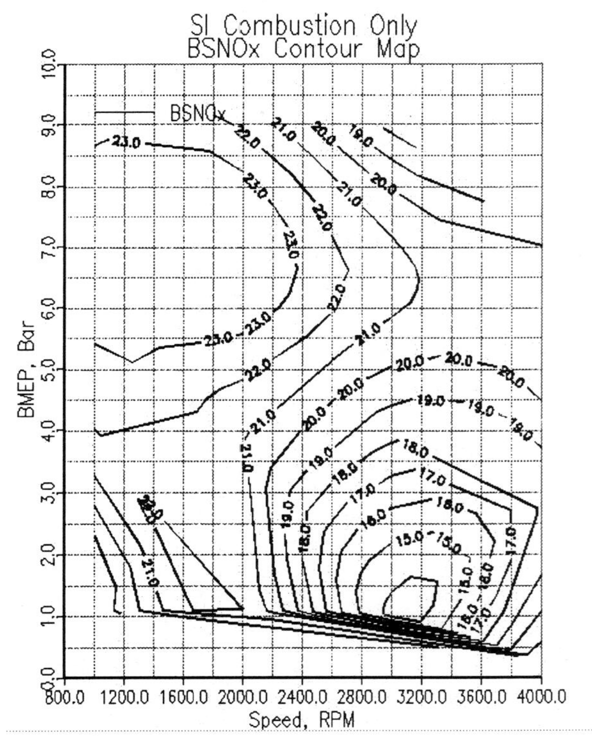

Figure 13

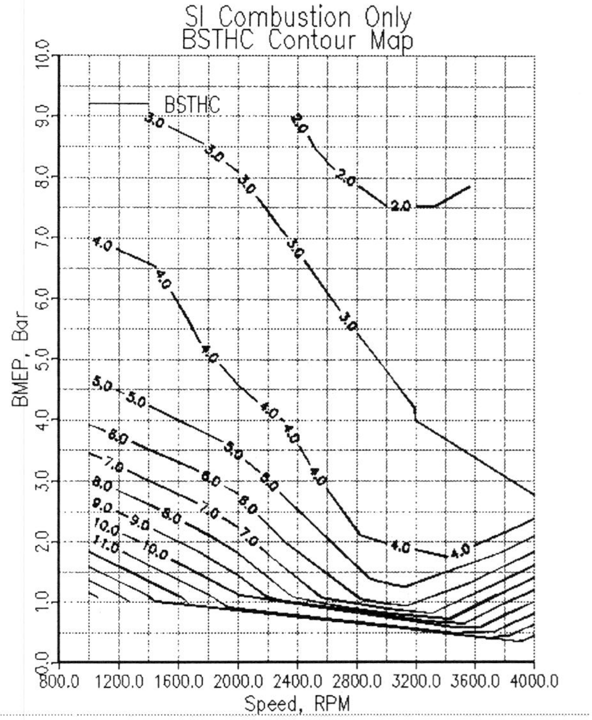

Figure 14

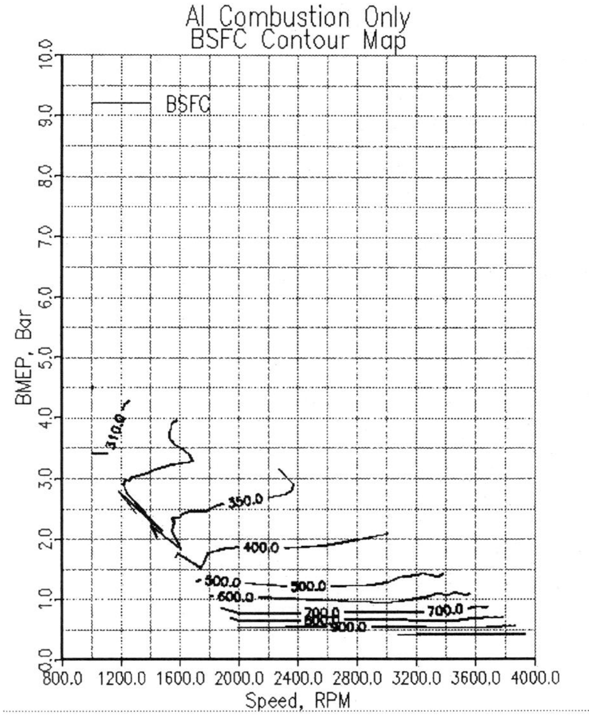

Figure 16

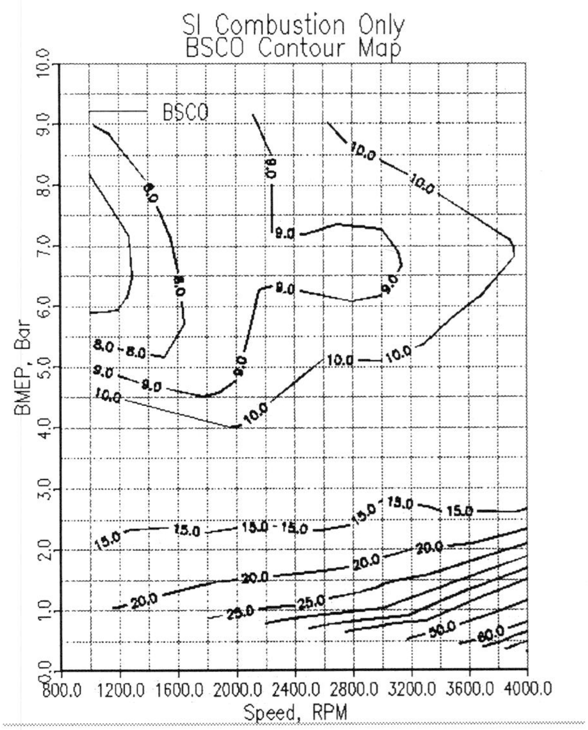

Figure 15

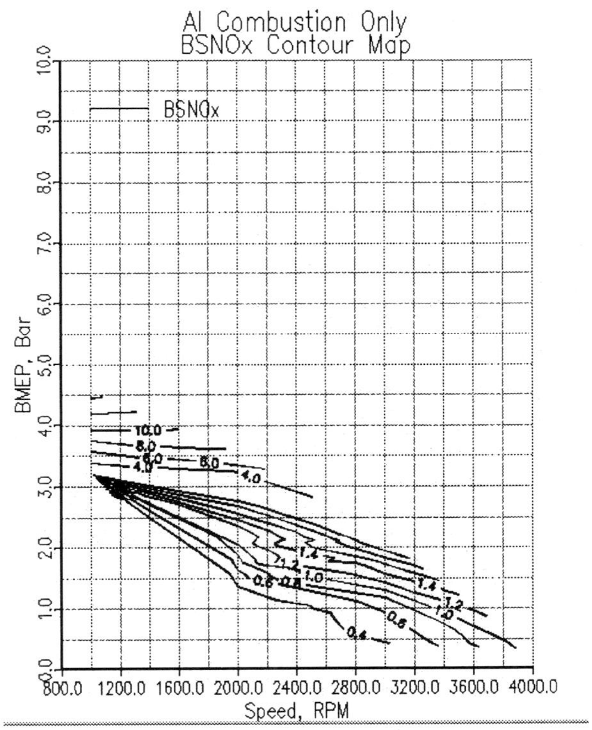

Figure 17

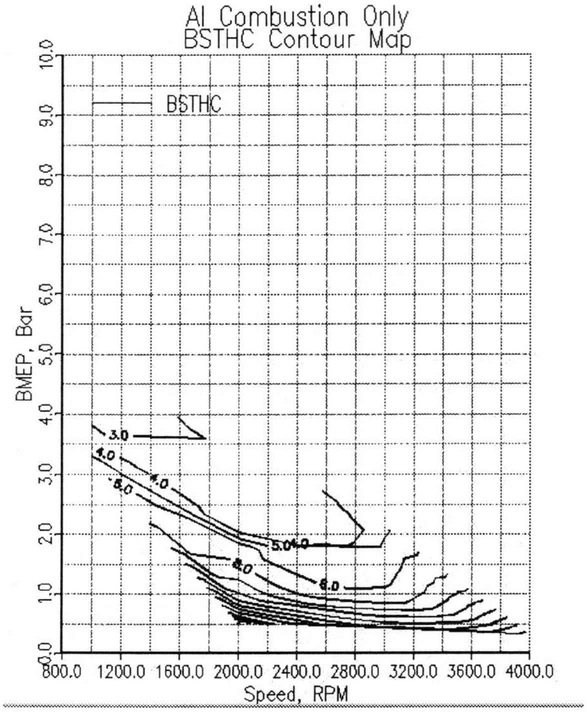

Figure 18

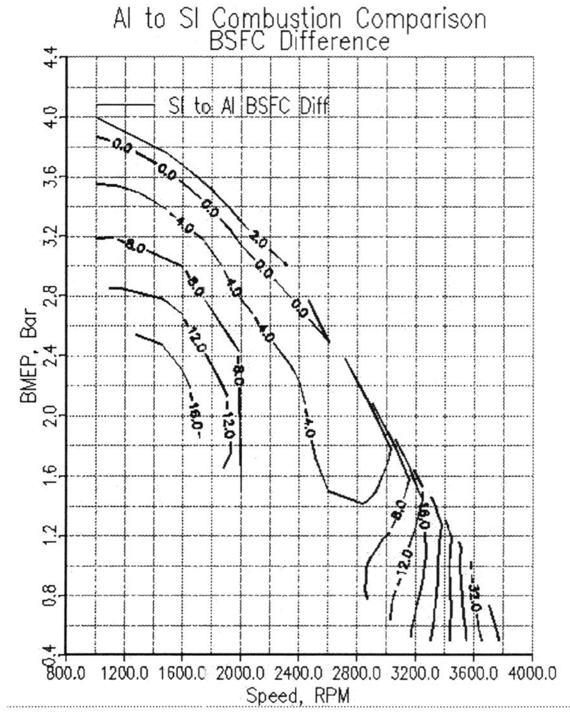

Figure 20

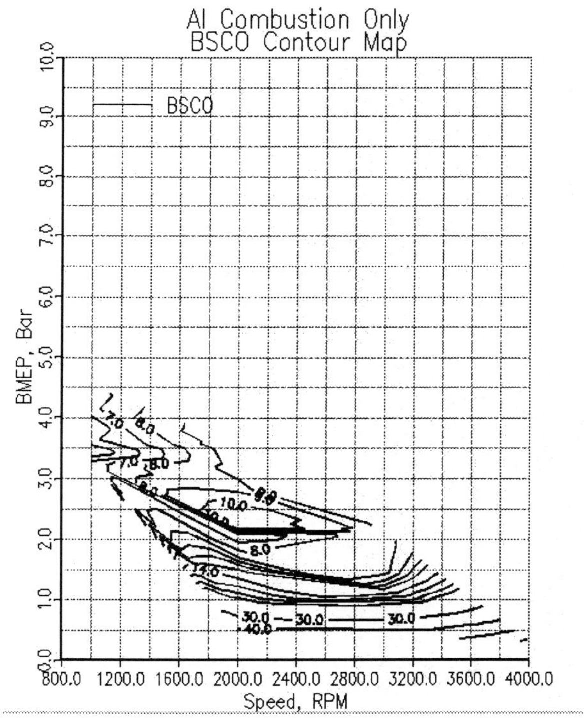

Figure 19

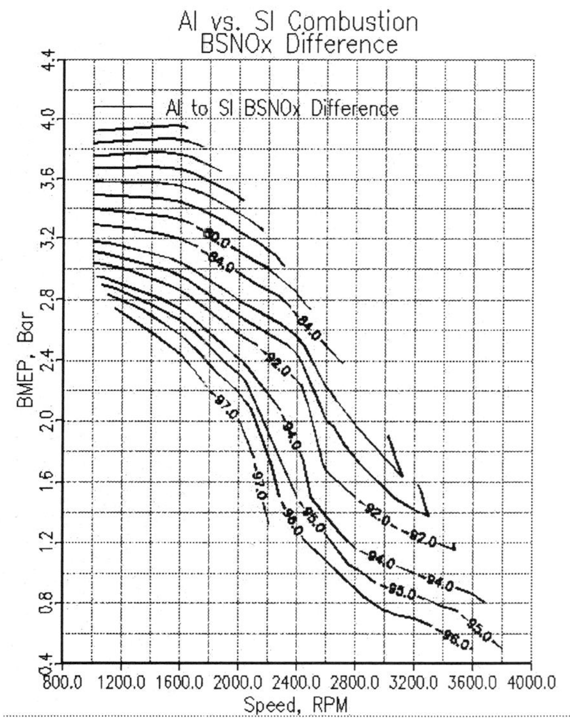

Figure 21

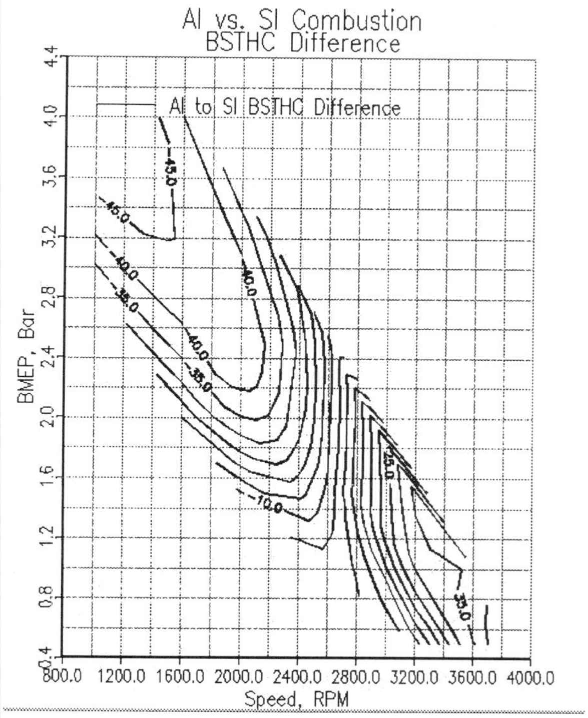

Figure 22

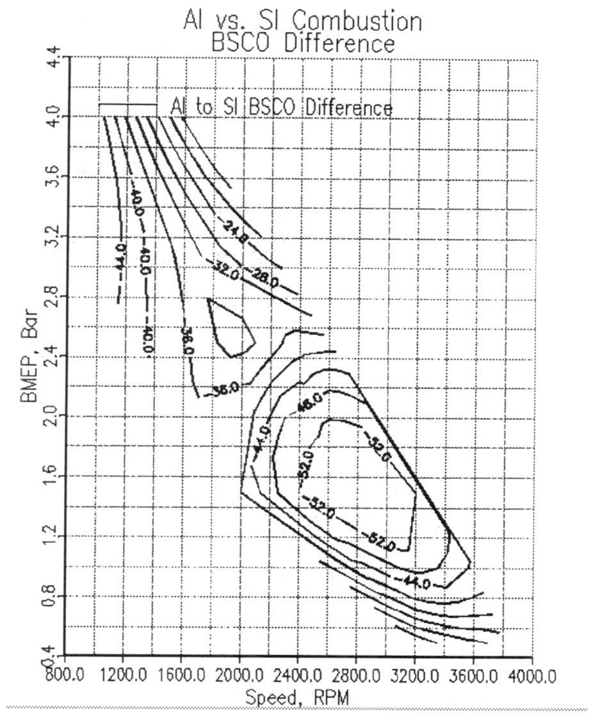

Figure 23

Figure 24: Fully optical single cylinder research engine used to capture the images shown below in Figure 25.

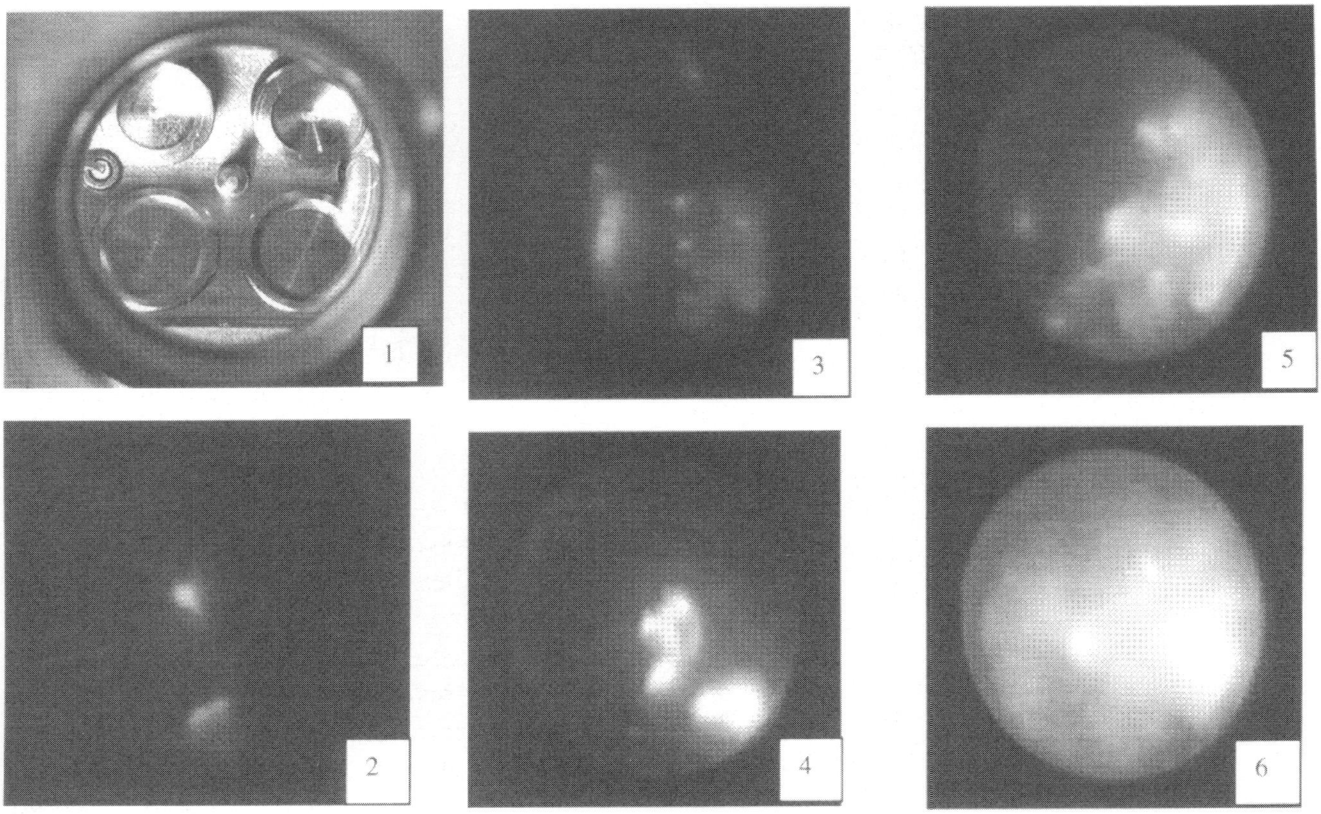

Figure 25: Images depicting CAI using the early exhaust valve closing method described earlier. In the examples above, a view though the optical piston can be seen (top left) together with multiple ignition points observed through a typical combustion cycle.

DISCUSSION

Feature	Research System	Production System
Maximum speed (rpm)	5000	8000
Soft touch down	Yes	Yes
Variable lift	Yes, up to 15mm	Yes, up to 10mm Gasoline 15mm Diesel
Variable timing	Yes	Yes
Closed loop position	Yes	Yes
Diesel and gasoline	Both	Both
Valve spring	No	Yes

Table 1 AVT system specification.

Bore	80.5mm
Stroke	88.2mm
Compression Ratio	10.5:1
Max speed	4000rpm
Number of valves	4
Fuel Injection	Port Injection
Fuel	Gasoline

Table 2 Single cylinder specification.

Feature	Intake	Exhaust
Lift	8.5mm	8.0mm
Duration	230cam deg	224cam deg
Max Open position	100 deg ATDC	104 deg BTDC

Table 3. Baseline valve timing.

CAI COMBUSTION IMAGES AND ADVANCED COMBUSTION

The data in Figure 25, shows the view through the sapphire piston crown looking at the inlet and exhaust valves. A spark plug can be seen to the far left of the image and a pressure transducer to the far right. In the centre is a blank (plug) where a direct injector can be fitted dependent upon experimental requirements. In this data set conventional port-fuel injection is utilised. Images are collected using a static mirror placed at 45° to the base of the piston crown and situated in the single cylinder engines crankcase-extension as shown in Figure 24.

The images in Figure 25 show CAI using the early exhaust valve closing strategy as described in Figures 8 and 9. The images are captured between ~5-15° crank angle after top dead centre (ATDC). This corresponds to a total volume of approximately 43cm^3 at a compression ratio of 10.5:1. Similar results to those obtained in Figure 25 are obtained using double opening of the exhaust valves as described earlier, thus demonstrating that FVVT's provide the essential tool through which advanced combustion research may be made possible.

The fact that similar combustion images are obtained using both methods demonstrates important similarities in the combustion type, despite two very different approaches. Additionally, since both methods produce CAI, this may give some mechanistic insight into the detail of CAI combustion. The question of whether CAI is promoted by chemical species (present in the exhaust gas or formed in fuel break-up) or directly from the combined effects of pressure and temperature remains unanswered. This question of mechanism has been discussed elsewhere [19] and is currently being investigated.

It is immediately clear from the images shown in Figure 25, that multiple ignition sites form during CAI combustion. From these initial sites, CAI can be seen to proceed via separate regions throughout the combustion chamber. It is also noticeable that, post-initial ignition site formation, numerous additional combustion sites form throughout the fuel-air/exhaust gas mixture. This implies that a compression-timing window (pressure and temperature derived) exists throughout which CAI is possible and that the formation of multiple sites suggests that local conditions *in-cylinder* govern initiation and propagation of CAI combustion. It is important to determine what are the benefits of such an advanced form of combustion and how CAI and SI need to be harmonised (perhaps even homogeneous charge compression ignition (HCCI)) to form hybridised forms of combustion. In such a way, the full benefits of CAI may be realised in terms of fuel economy and pollutant reduction.

It is well known that in conventional spark ignition, once a spark has been initiated, the resultant combustion kernel is strongly dependent on local conditions adjacent to the spark itself. Local (1) air:fuel effects, (2) air motion, (3) uniformity/density of spark and (4) initial chemical reaction kinetics will all form crucial parts describing the reproducibility of the initial combustion. Since there are a number of important factors influencing reproducibility (random and systematic), the net result is a cycle-to-cycle variation generated in the initial combustion kernel with a subsequent effect on flame propagation.

Overall, this leads to a reduction in the reproducibility of combustion events with a direct consequence that in order to compensate for this, and thus gain a stable more reproducible combustion, deliberate over-fueling must be employed. In short, we use more fuel than is desired with a corresponding output of this action being a net increase in engine-out emissions which must then be rectified using after-treatment based catalyst technologies. An improved approach would be to gain much greater control of the combustion event itself, therefore effecting greater cycle-to-cycle reproducibility with a resultant that less fuel is required and so fewer pollutants are created. This is where CAI has distinct advantages.

VALVE STRATEGIES AND ADVANCED COMBUSTION OPERATION

Figure 26 shows the trend in control technologies from carburetors through single and multi-point injection to direct injection and then to fully variable valve systems. This is a clear trend in improved control over the fluid handling processes associated with the combustion event. All leading towards improved combustion and reduced fuel consumption and emissions. Traditionally it has been fuel control that has been gradually increased, but today we are seeing greater and greater control being excerised over the gas exchange process required for combustion. This has lead to the introduction of more sophisticated valve actuation systems and will lead eventually to fully variable valve actuation such as the AVT system.

Use of a FVVT on a single cylinder engine has demonstrated the potential for improving fuel efficiency and reducing emissions from a gasoline SI internal combustion engine.
In detail, three different throttleless valve strategies have been shown in this paper and each can be used to achieve load control of the engine by varying the timing of individual engine valves.

We have shown an "Early Inlet Valve Closing" strategy, which gives the possibilities of achieving true throttleless SI combustion without the need for stratified combustion. Also, two alternative valve strategies are shown which both achieve load

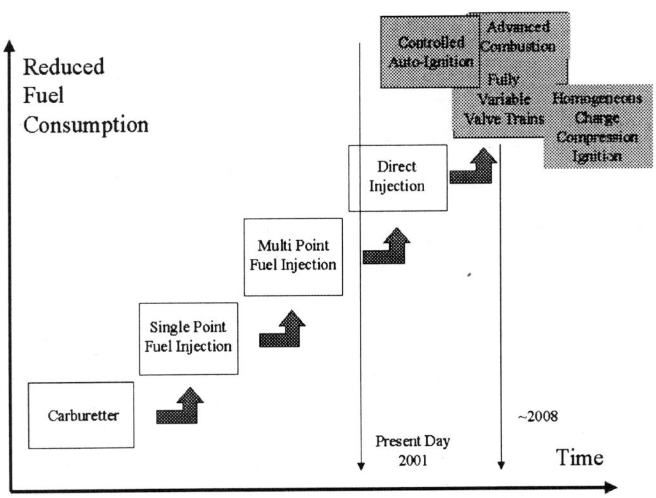

Figure 26: Improved fluid control.

control and CAI combustion. All three valve-operating strategies can be varied to achieve load control without the need for a throttle. The FVVT generates the ability to move either progressively or digitally from valve timing to valve timing to control the load of the engine and thus control the operating mode of the combustion process. This flexibility allows engine calibration and development engineers to select the most appropriate combustion strategy to deliver the required power output whilst significantly reducing emissions and improving fuel economy compared to standard SI engines. Using the initial results from the mapping of the single cylinder shown in Figures 12 to 23, it is possible to compare combustion operating strategies and identify areas of operation which are preferred operating points and deliver benefits of up to 95% reduction in NOx and a 40% reduction in THC. Likewise areas of none-preferred operation can be avoided, such as areas on the high load limit of CAI where uncontrolled detonation is more likely to occur.

CONCLUSIONS

The work presented in this paper shows the extent of the operating maps for the CAI combustion process when running in a stoichiometric air fuel ratio. The area of CAI operation can be extended by further modification of the in-cylinder conditions by modifying the air fuel ratio and running the engine lean, this work is currently ongoing and will be reported in subsequent papers.

This paper has presented a workable area of CAI which has benefits of significant emission reduction, (80%-97% NOx and 35%-45% THC) and also significant fuel economy reduction due to elimination of pumping losses and enhanced reproducibility of the combustion event. Also shown are valve timing strategies, which may be used to achieve both throttleless CAI and throttleless SI. The FVVT system demonstrates the ability to change valve timing from cycle-to-cycle and allows either a gradual migration from SI to CAI or a rapid switch from one combustion mode to the other.

The authors believe that it is precisely this type of valve control system, with its true flexibility, that will allow engineers to exploit throttleless SI and CAI and will ultimately lead to a new generation of advanced engines required to meet future customer, fuel economy and emissions legislation demands.

ACKNOWLEDGMENTS

We would like to acknowledge Ted Copland and Clem Sayers for their help with this work.

CONTACT

Further information may requested from JAllen@lotuscars.co.uk or DLaw@lotuscars.co.uk.

REFERENCES

1. N.Wilson, J.Allen, G.Muddel, Future Valve Train Systems to Reduce Engine Performance Compromises, FISITA Conference 1992.
2. N.Wilson, J.Allen, G.Muddel, Lotus Variable Valve Train Systems, Haus der Technik, Essen, 1992.
3. J.Allen, N.Wilson, I.Mech.E Paper C462/34/178, 1993.
4. P.Langden, R.Cosfeld, A.Grundno, K.Reif, Electromechanical Valve Actuation: A Basis

for Future Gasoline Engine Concepts, BMW Group, 21st Internationales Weiner Motoren-symposium 2000, p208-225.

5. N.B.Kaahaaina, A.J.Simon, P.A.Caton and C.F.Edwards, Use of Dynamic Valving to Achieve Residual Affected Combustion, SAE 2001-01-0549.
6. S.Onishi, S.H.Jo, K.Shoda, P.D.Jo and S.Kato, Active Thermo Atmosphere Combustion (ATAC) – A New Combustion Process for Internal Combustion Engines, SAE790501.
7. M.Noguchi, Y.Tanaka, T.Tanaka and Y.Takeuchi, A Study on Gasoline Engine Combustion by Observation of Intermediate Reactive Products During Combustion, SAE790840.
8. Y.Ishibashi and Y.Tsishima, A New Generation of Engines for the Future, P.Duret (Editor) and Editions Technip, Paris 1993, p113-124.
9. Y.Ishibashi, Basic Understanding of Activated Radical Combustion and its 2-Stroke Engine Application and Benefits, SAE 2000-01-1836.
10. N.Iida, Combustion Analysis of Methanol Fuelled Active Thermo Atmosphere Combustion (ATAC) Engine Using Spectroscopic Observation, SAE940684.
11. P.Duret and S.Venturi, Automotive Calibration of the IAPAC Fluid Dynamically Controlled 2-Stroke Com-bustion Process, SAE960363.
12. N.Iida and T.Igarashi, Auto-ignition and Combustion of N-Butane and DNE/Air Mixtures in a Homogeneous Charge Compression Ignition Engine, SAE200-01-1832.
13. E.Esterlingot, P.Guibert, J.Lavy and S.Raux, Thermodynamical and Optical Analyses of Controlled Auto Ignition Combustion in 2-Stroke Engines, SAE972098.
14. M.Kumada, H.Shimodaira, K.Yoshida, H.Shoji and A.Saima, Proc. 15th Int.Comb.Eng.Symp., Seoul Korea, July 13-16th, 1999, p395-400.
15. P.Duret, J-C.Dabadie, J.Lavy, J.Allen, D.Blundell, J.Oscarsson, G.Emanuelsson, M.Perotti, R.Kenny and G.Cunningham, The Air Assisted Direct Injection ELEVATE Automotive Engine Combustion System, SAE Fuels and Lubes, SAE FL487.
16. P.Duret, J-C.Dabadie, J.Lavy, J.Allen, D.Blundell, J.Oscarsson, G.Emanuelsson, M.Perotti, R.Kenny and G.Cunningham, SAE Fuels and Lubes, Innovative Design and Simulation Study of the ELEVATE Automotive Engine Project, SAE Congress, Paris, 2000.
17. M.Stockinger, H.Schapertons and P.Kuhlman, MTZ 53(1992)2, p80-85 (in German).
18. Lavy *et al.* (14 authors), Innovative Ultra Low NOx Controlled Auto Ignition Combustion Process for Gasoline Engines: the 4-SPACE Project, SAE 2000-01-1837
19. D.Law, J.Allen, D.Kemp and P.Williams, 4-Stroke Controlled Auto Ignition Investigations Using a Single Cylinder Engine with Lotus Active Valve Train (AVT), Proceedings of the 21st Century Emissions Technology Conference, I.Mech.E, London, 4-6th December 2000, Paper C588/006/2000.
20. D.Law, D.Kemp, J.Allen, G.Kirkpatrick and T.Copland, Controlled Combustion in an IC-Engine with a Fully Variable Valve Train, SAE 2001-01-0251.
21. P.M.Najt and D.E.Foster, Compression Ignited Homogeneous Charge Combustion, SAE 830264.
22. R.H.Thring, Homogeneous Charge Compression Ignition (HCCI) Engines, SAE 892068.
23. T.Aoyama, Y.Hattori, J.Mzuta and Y.Sato, An Experimental Study on Premixed Charge Compression Ignition Gasoline Engine, SAE 960081.
24. A.Oakley, H.Zhao, N.Ladommatos and T.Ma, Experimental Studies on Controlled Auto-Ignition (CAI) Combustion of Gasoline in a 4-Stroke Engine, SAE 2001-01-1030.

A New Generation of Engine Combustion Processes for the Future?
© P. Duret (Editor) and Editions Technip, Paris, 2001, pp. 101 - 114
27 rue Ginoux, 75015 Paris

CONTROLLED AUTO-IGNITION (CAI): A NEW HIGHLY EFFICIENT AND NEAR-ZERO NOx EMISSIONS COMBUSTION PROCESS FOR GASOLINE ENGINE APPLICATION

Jacques Lavy , Jean-Charles Dabadie, Pierre Duret, Christian Angelberger, Jean-François Le Coz, Jérôme Cherel
IFP, France

ABSTRACT

New engine combustion processes such as diesel Homogeneous Charge Compression Ignition (HCCI) and gasoline Controlled Auto-Ignition (CAI) are seen among the most promising ways to meet the environment challenges of the future powertrains used in transportation, which will have to target lower and lower CO2 emissions and fuel consumption (ACEA commitments) with associated robust ultra low pollutant emissions.

In this paper, it will be shown how by using appropriate control strategies of the internal EGR charge, it is possible to obtain the auto-ignition of a lean mixture and to control it (in terms of ignition timing, heat release rate, cylinder pressure increase, ...) in an existing automotive gasoline 4-stroke engine with conventional compression ratio and without requiring heating of the intake air charge.

In such controlled conditions, and compared to the conventional Spark Ignition (SI) combustion process, the newly-developed lean burn CAI combustion process allows to achieve part load fuel consumption improvements in the range of 15% to 20% combined with near zero NOx emissions ("one digit" ppm of NOx emissions is usual in some conditions).

By allowing to run a gasoline engine in lean burn mode without requiring any fuel-sensitive DeNOx aftertreatment technology, the CAI combustion process further increases the already known Variable Valve Actuation (VVA) technologies benefits in terms of fuel efficiency and low CO2 without needing any other engine modification or redesign. It is therefore expected that the coming new generation of VVA technologies and associated engine management system, will make possible the introduction of the CAI combustion process (in at least some operating range) in the automotive gasoline engines of the short to medium term future.

INTRODUCTION

The stoichiometric gasoline four-stroke with 3-way catalyst for passenger car has the potential (with additional strategies or device such electric heated catalyst,...) to reach ultra low emissions levels such as the Euro stages 4 and 5 for years 2005 and 2008. Nevertheless, it suffers of dramatically poorer efficiency compared to lean burn concepts.

On the other side, diesel engines as well as gasoline direct injection engines are promised to an increased part of the transport market if their NOx and particulate emissions are reduced. Today, adequate after-treatments for NOx and PM traps are under industrialization with still some concerns about fuel economy, robustness, sensitivity to fuel sulphur and cost.

New combustion processes are investigated for their potential to reduce towards very low level the production of NOx. They constitute alternative or complementary answers to the sophisticated and complex after-treatments strategy. Homogeneous Charge Compression Ignition (HCCI) and a number of acronyms designate these combustion. Their principle consists in preparing an highly diluted mixture, in achieving its simultaneous ignition in the whole space of the combustion chamber and in precisely controlling such combustion for best performance in terms of efficiency and NOx emissions (timing and heat release rate during the cycle). Thanks to the thermo-chemistry of the charge, the thermal NOx formation and the soot production are in principle much less than with the diffusion flame typical of the conventional diesel or with the propagation flame typical of the conventional spark ignition engine [1].

The purpose of this paper is to show how the IFP-developed gasoline Controlled Auto-Ignition (CAI™) seems to be a promising way to meet the environment challenges of the future powertrains used in transportation, which will have to target lower and lower CO2 emissions and fuel consumption (ACEA commitments) with associated robust ultra low pollutant emissions.

To qualify the approach of gasoline auto-ignition presented in this paper, the terms “gasoline Controlled Auto-Ignition” or “gasoline CAI™” are generally preferred to the terms “gasoline Homogeneous Charge Compression Ignition” or “gasoline HCCI”. The reason is that in this approach, it is possible to auto-ignite the fresh charge in a 4-stroke gasoline engine without using intake air heating neither increased compression ratio. It means that the same engine is also able, without engine design modification, to move from gasoline CAI at part load to SI (Spark Ignition) at full load for maintained performance. IFP was probably the first to publish this innovative approach based on the use of a high quantity of stratified internal EGR in a 4-stroke engine [2] as it can be inherently achieved in 2-stroke engines [3]. This approach has been clearly detailed in previous papers [3-9]. Because this gasoline CAI is not homogeneous (at least between the internal EGR and the fresh charge), the “H” letter of “HCCI” becomes therefore not appropriate. This is why the term "gasoline CAI" is here much preferred to "gasoline HCCI" [10].

In this sense, this approach clearly differs to other gasoline HCCI approach, which proposes really “Homogeneous” combustion based on high compression ratio or intake air charge heating [11-13] which makes more difficult the transition to the full load operation without knock.

HOW TO GET GASOLINE AUTO-IGNITION IN A 4-STROKE ENGINE?

In the first section of this paper, we will describe the approach and methodology adopted by IFP in the development of the gasoline Controlled Auto-Ignition combustion process. It is based on the use of scientific tools combined with newly available innovative engine technologies and control strategies. It has been partly develop within the 4-SPACE (4-Stroke Powered gasoline Auto-ignition Controlled combustion Engine) European Consortium project [6, 8-10, 14, 15].

THE RESEARCH APPROACH AND CORRESPONDING METHODOLOGY ADOPTED

An overview of the research approach which was necessary to design and test different concepts to use controlled auto-ignition in 4-stroke engines is described in this section.

The project flow diagram here below (Figure 1) summarizes the main tasks undertaken within the 4-SPACE project. The interdependence between these tasks shows the methodology adopted which was necessary to go up to the in-vehicle predicted evaluation of the 4-SPACE concept applied to a 4-stroke multi-cylinder automotive engine [9].

FROM THE ANALYSIS OF THE CONDITIONS OF AUTO-IGNITION IN A 2-STROKE ENGINE

As shown by Figure 1, the IFP contribution to the 4-SPACE research project started by the analysis of controlled auto-ignition in a 2-stroke engine [8,14] since such type of engines can be designed to be able to run in this combustion mode on a wide operating range, due to its specific high internal EGR rate at light load [3]. In this first step, the objectives were to well understand the in-cylinder conditions which promote auto-ignition such as, for example, the EGR rate, the in-cylinder pressure, the in-cylinder gas temperature gradient, the stratification level between the fresh charge and the residual gas,... Test bench measurements, 3D CFD modeling and in-cylinder visualization were performed to get data collection leading to a precise understanding of these 2-stroke in-cylinder conditions when Controlled Auto-Ignition occurs.

Figure 2 illustrates the linear relationship between the typical residual gases rate and their temperature at auto-ignition onset in various CAI running conditions. We can see the highest residual gas quantity for very light load running conditions. In such condition, since the energy released by the combustion is rather low (small quantity of fuel is burnt by cycle), exhaust gases temperature is low and residual gases temperature prior to auto-ignition level is therefore low.

As far as the engine load and IMEP increase near medium load, the residual gases rate decreases (more fresh charge must be kept in the cylinder to get higher energy from the combustion) but their temperature level increases (due to higher released energy). Therefore, for higher load conditions, internal EGR rate decreases and CAI process disappears even though residual gases temperature continues to increase.

The CFD modeling approach was also very useful to study and better understand the 2-stroke gasoline CAI combustion process. An example of application to the analysis of the mixing process and temperature fields in a gasoline CAI 2-stroke engine is shown in Figure 3. In this example, fully described in a previous papers [8], the overall temperature level is almost similar between the SI and the CAI case. The only difference is that to obtain CAI combustion, the transfer ducts are partially throttled to smooth the fresh charge introduction, which reduces the mixing between the fresh charge and the inherent internal EGR. This increased stratification between fresh charge and internal EGR is clearly shown by the increased temperature gradient obtained in the CAI CFD calculation.

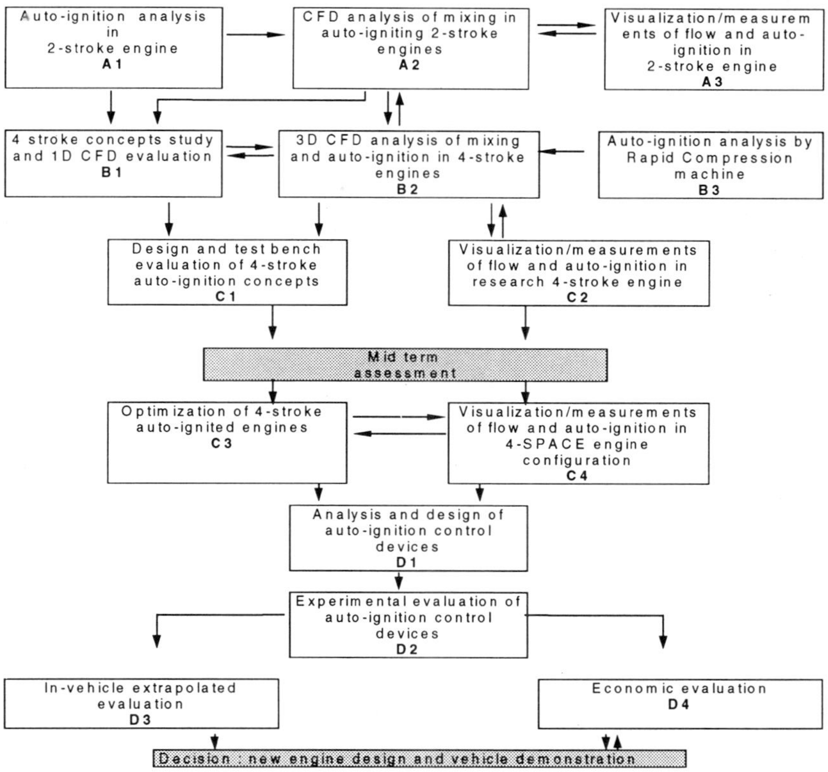

Fig. 1: 4-SPACE project flow diagram [9]

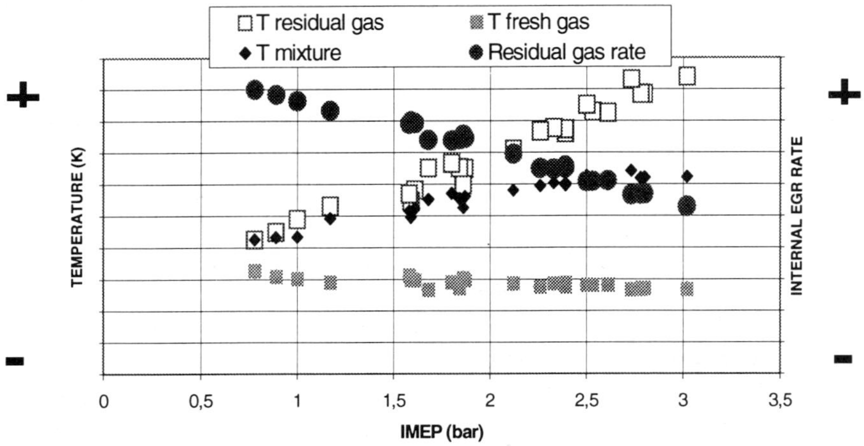

Fig. 2: Temperatures and residual gas rate versus IMEP in a CAI 2-stroke engine

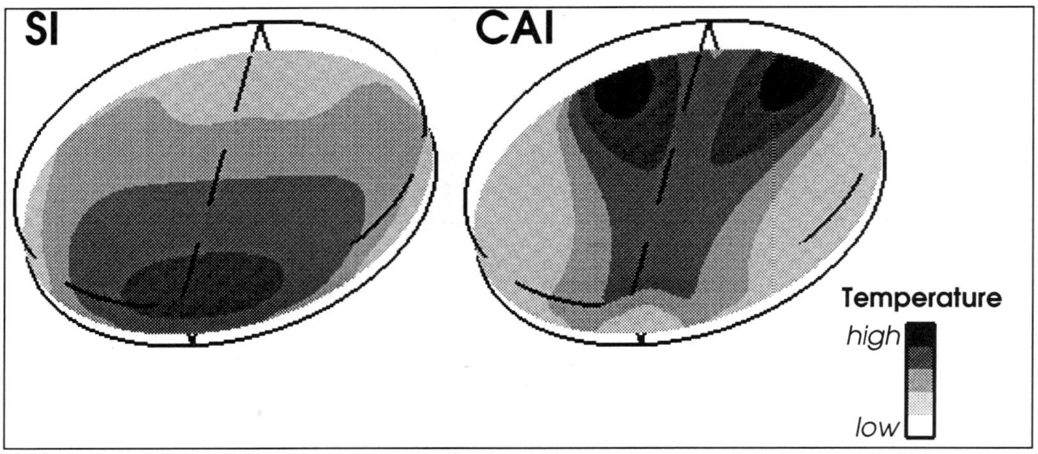

Fig. 3 : Compared SI and CAI temperature fields in the combustion chamber at 10 degree CA BTDC

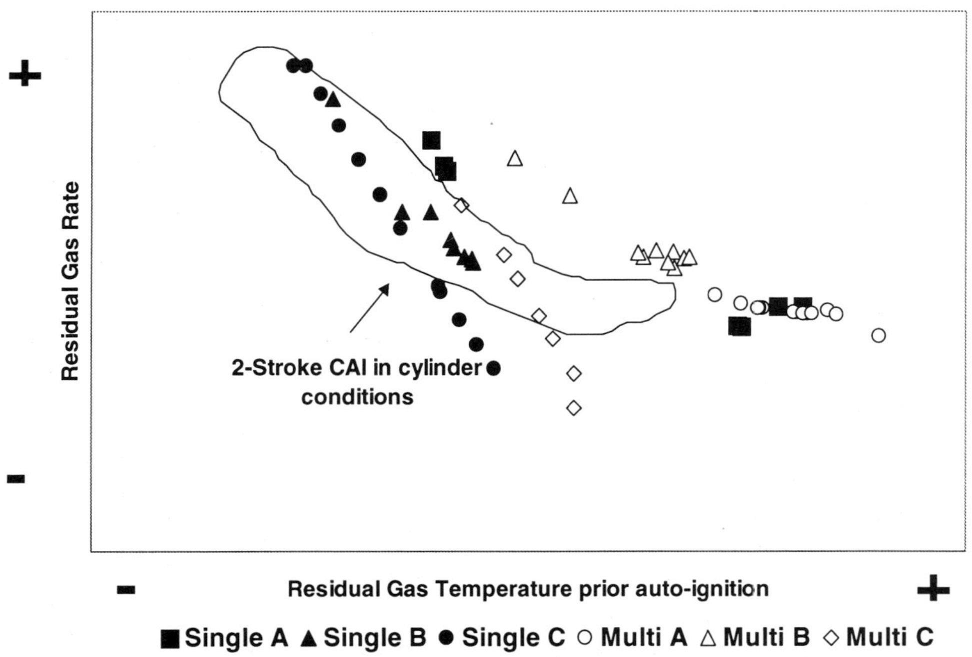

Fig. 4: "Global" in-cylinder conditions of EGR versus temperature - 1D calculations results of different CAI 4-stroke engine concepts compared to 2-stroke experimental data

TO THE FIRST APPLICATION OF THE CAI COMBUSTION IN A 4-STROKE ENGINE

Then, in a second step of the 4-SPACE project, the work was devoted to the study and analysis of controlled auto-ignition concepts for 4-stroke engines in order to try to reproduce conditions as similar as possible to the in-cylinder 2-stroke engine conditions already analyzed. It was expected that this transfer of 2-stroke conditions to the 4-stroke engine application could be done with the help of specific control devices such as specific valve event configuration allowing to reproduce the 2-stroke internal EGR charge in the case of the IFP-developed concept. 1D and 3D CFD helped us to evaluate the different configurations potential.

1D calculations have been first extensively performed to validate the potential of potential 4-stroke CAI concepts in terms of "global" conditions such as concentration of hot internal residual gas and temperature level as shown in Figure 4. Some of them are related to so-called "single-cylinder" engine approach (referenced Single X on the figure) where each cylinder can run independently to the others. In most of these concepts, the main engine design modifications consist of special valve opening characteristics (timing and lift) allowing to trap a high amount of hot residual gases [16]. The others concepts consider specific multi-cylinder arrangements (referenced Multi Y) with, for the same purpose, some connections between cylinders [17].

Their potential in terms of in-cylinder conditions of EGR is compared to experimental 2-stroke engine data in CAI combustion mode (corresponding to the previous results of Figure 2). The fact that for some simulated configurations, the results are located in the most favorable conditions range (that is both internal EGR rate and EGR temperature similar or higher than CAI 2-stroke engine data) indicates that the necessary conditions of internal EGR quantity and temperature can be met. This indication was nevertheless not considered as sufficient to be sure that CAI combustion should happen in an actual engine since these 1D calculations do not take into account the 3D mixing aspects which modify the "local" conditions as identified from 3D CFD analysis in 2-stroke engine (Figure 3). Therefore, in a following step, the most promising 4-stroke CAI concepts from 1D calculations were calculated in 3D conditions [9] and implemented in an experimental engine for development and assessment.

Then the next Figure 5 compares similar 3D CFD calculations between 2-stroke and 4-stroke conditions of CAI combustion. It can be seen less stratification in the 4-stroke case where time for mixing is probably longer due to the cycle frequency difference.

Nevertheless, the fact that the 4-stroke effective compression ratio is higher than the 2-stroke one probably compensates the higher 4-stroke mixing effect.

Therefore, in this section of the paper, the research and technical approach and methodology used to develop the CAI combustion process have been shown. This methodology was concluded by successful 4-stroke engine operation in not only auto-ignition mode, but really in "Controlled" Auto-Ignition or CAI as it will be described in the next section of the paper.

Before to move to the next section, it is nevertheless interesting to show how a complementary scientific technique, such as optical diagnostics in an operating engine, can help in the understanding of this new combustion process. Detailed measurements and results of CAI combustion in a transparent engine will be presented in a future paper. The figure 6 presents preliminary results of the fuel concentration visualized by laser induced fluorescence in gasoline CAI combustion for 3 different cycles and at the same crankangle (4 deg. ATDC). The dark area corresponds to the area of already burned gases. It can be clearly seen that the locations of burned gases areas are not homogeneous within the same cycle, and also from a cycle to another cycle.

On the contrary, in gasoline CAI, the cylinder pressure trace as well as the cylinder IMEP (not shown here) are extremely stable from one cycle to another cycle. Therefore, such sophisticated optical diagnostics represent precious tools to progress in the understanding of this CAI combustion and to support its development up to practical application.

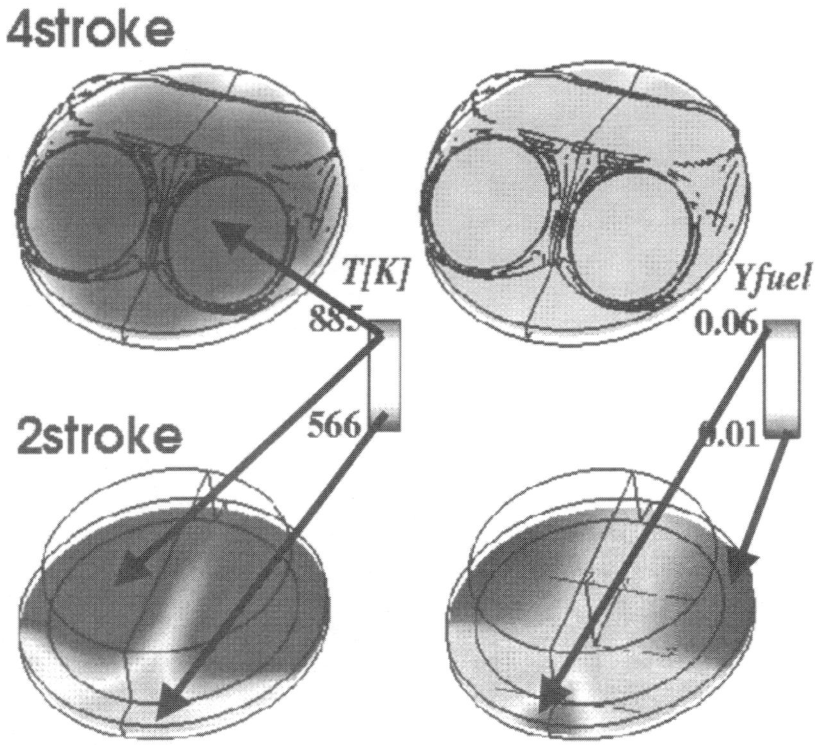

Fig. 5 : Compared 3D CFD results between 2-stroke and 4-stroke CAI combustion: temperature fields and fuel concentration in the combustion chamber at 10 deg. CA BTDC

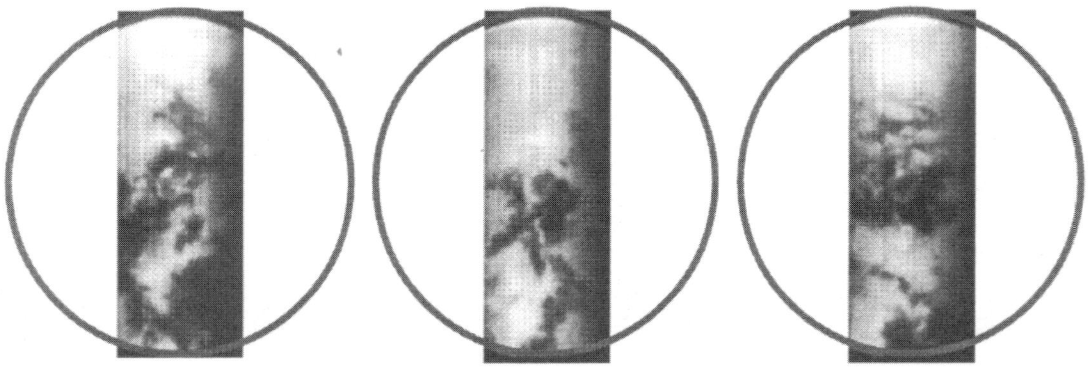

Fig. 6: Visualization of fuel concentration by laser induced fluorescence in a gasoline CAI engine - the dark zones correspond to burned gases (4 deg. CA ATDC - 1500 rpm – lambda 1.12 – CR=10)

HOW TO CONTROL THE GASOLINE AUTO-IGNITION IN A 4-STROKE ENGINE?

From the steps used and described in the previous section of the paper, it has then been possible to experimentally develop and test some of the gasoline auto-ignition concepts that were numerically studied. The first experimental results have already been published [6-8].

They show that by reproducing the 2-stroke internal EGR charge strategy, it is possible to achieve the main breakthrough of 4-stroke auto-ignition with low engine compression ratio (between 10 to 12.4 in all the results presented in this paper) and without requiring intake air heating. In addition, and as expected from the 2-stroke experience, such combustion process can achieve near zero NOx emissions (up to 200 to 300 times reduction compared to stoichiometric SI operation) combined with 10 to 20% fuel consumption reduction according to the engine operating conditions. Bringing a solution to overcome the NOx-fuel consumption dilemma, the CAI combustion can therefore represent a less sulphur sensitive and emissions robust alternative strategy to the stratified DI strategy with sophisticated DeNOx aftertreatment.

Thanks to its near zero NOx emissions level, lean operation can be used in CAI mode for best efficiency. Switching to more conventional stoichiometric SI mode remains nevertheless necessary at the higher loads which are outside the CAI range.

Based on these considerations, the main further development work on CAI has therefore to be concentrated on two main issues:

- still research and develop any solutions and EMS control strategies that allow to precisely control the CAI combustion timing and heat release rate, and its transition to the SI combustion at higher loads,
- extend the CAI range to the widest range as possible in terms of engine speed and load in order to take more benefit of the fuel consumption and CO2 emissions advantages of this combustion technology.

CONTROLLED AUTO-IGNITION RANGE

The next Figure 7 shows typical engine operating range already achieved in CAI mode for two different air charging strategies, one being achieved with naturally aspirated intake air and the other one with boosted intake air.

In this figure, three reference points are indicated. They will be the subject of the experimental results presented in this paper. The NA-CAI range of Figure 7 is obtained with a special configuration of valve events (with fixed valves lift) designed and optimized around the reference point N°1. The Boosted-CAI range is obtained with another special valve events configuration and charge control strategy.

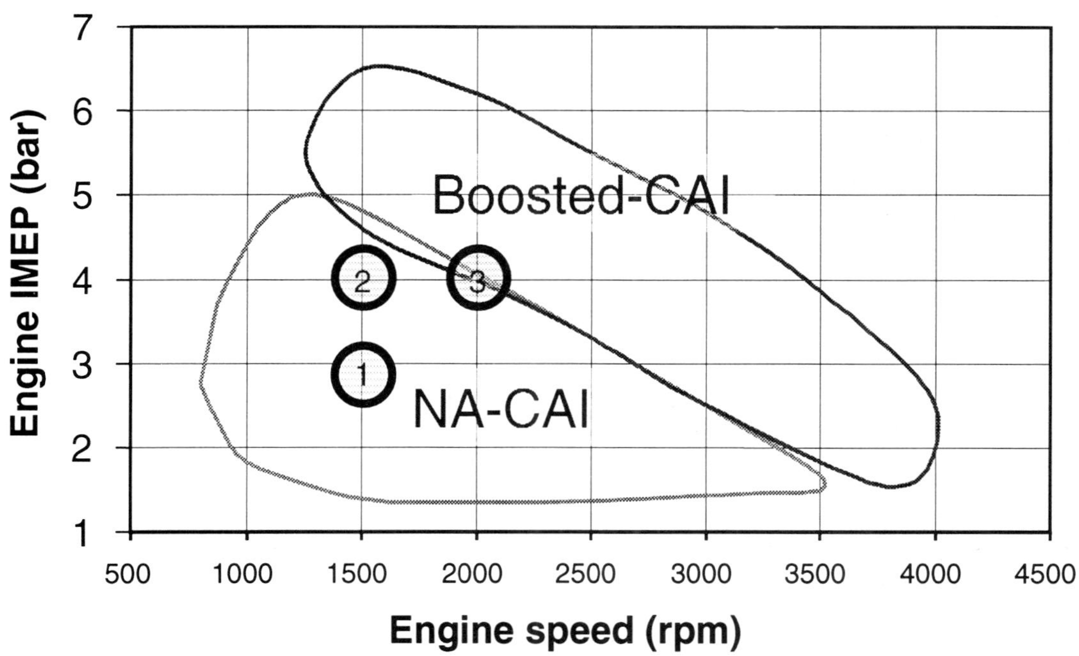

Fig. 7: CAI range and studied reference points for two different air charging strategies, one Naturally Aspirated (NA-CAI) and one with boosted intake pressure (Boosted-CAI) – (ambient intake temperature)

TYPICAL CONTROLLED AUTO-IGNITION PERFORMANCE

The next Figure 8 presents the results achieved in NA conditions when the valve lift configuration is optimized around the reference point N°1 of Figure 7 (ie 2.9 bar IMEP @ 1500 rpm).

In this figure, results are compared between SI and CAI combustion modes. The compression ratio is the same in both cases as well as all the engine design except the intake and exhaust valves opening characteristics. In both SI and CAI, the mixture is leaned from stoichiometry to lambda 1.7.

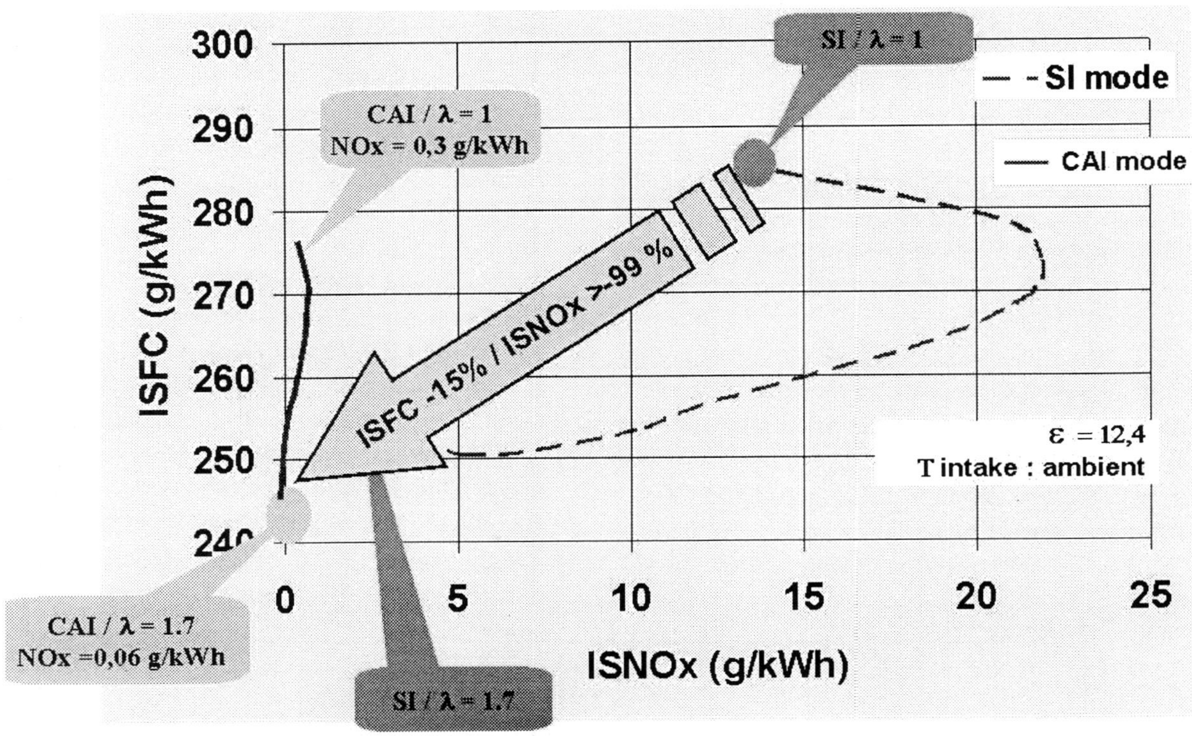

Fig. 8: Example of typical CAI effect on the Fuel Consumption (or CO2) / NOx trade-off during part load operation (Reference point N°1: 2.9 bar IMEP @ 1500 rpm)

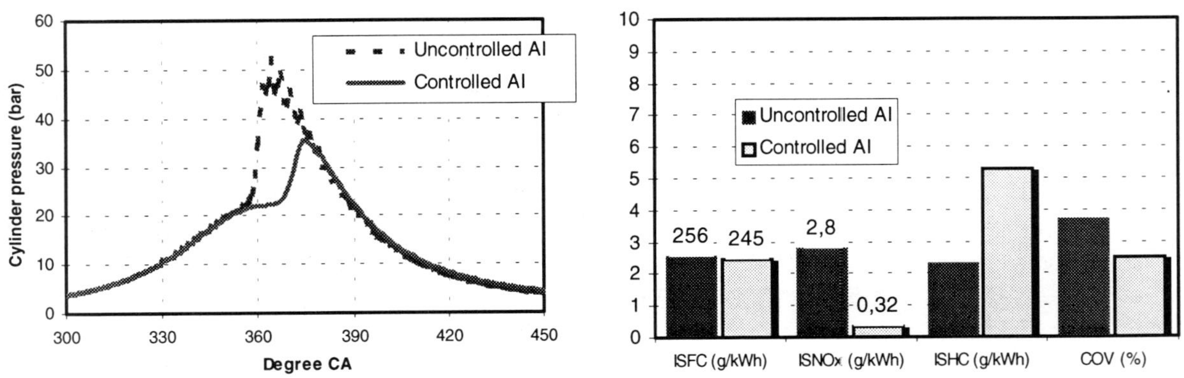

Fig. 9: Compared cylinder pressure traces, fuel consumption and emissions results in two different conditions of auto-ignition control strategy (Reference point N°2: 4 bar IMEP @ 1500 rpm)

In SI mode, the Indicated Specific Fuel Consumption (ISFC) decreases as usual with firstly an increase of NOx emissions, followed by a decrease to a minimum where the BSFC start to re-increase. At this point, the leanest limit is reached, the HC emissions (not shown here) increases significantly and ISFC is improved by about 12% compared to stoichiometric SI. Nevertheless, even if NOx emissions are reduced, they remain in the range of 3-5 g/kWh which is very much higher than the level required to meet Euro 4 legislation. This means that to take benefit of the lean burn fuel economy advantage, this SI lean solution would require a sophisticated and highly efficient DeNOx aftertreatment

In the CAI mode, stoichiometric ISFC is already 4% lower than in SI, and NOx are around 0.3-0.5 g/kWh which means about 40 times lower than in SI. The great interest of CAI is that the engine can run in lean operation (lambda 1.6-1.7) with NOx further reduced to minimum levels of 0.03-0.05 g/kWh (about 4 to 5 ppm) which are 250 to 300 times lower than stoichiometric SI, and with fuel consumption reduced by 15 % in this example (and slightly more in some other cases). This means that at this operating point, lean burn CAI combustion gives a NOx level easily compatible with the Euro 4 standards without aftertreatment and with significant fuel (and therefore CO2) economy compared to stoichiometric SI.

THE TRAPPED EGR CHARGE STRATEGY TO CONTROL THE CAI TIMING AND HEAT RELEASE

To get the impressive results obtained in the previous Figure 8, a precise control strategy of the concentration of trapped residual gases, of their respective temperature level and of their degree of mixing with the fresh air charge is required. This is obtained by combining special valve events (to control the trapped EGR quantity and the overall temperature level inside the combustion chamber) with internal fluid dynamic control of the mixing between the fresh charge and the EGR (to control the temperature gradient inside the combustion chamber).

The Reference point N°2 (4 bar IMEP @ 1500 rpm) of Figure 7 can firstly be obtained with the same valve event configuration of the Reference point N°1. In this case, it becomes located in the upper limit of the CAI range optimized around the Reference point N°1. The auto-ignition becomes too early (see Figures 9 and 10 "Uncontrolled AI") with a strong in-cylinder pressure rise and with increased NOx emissions, even if the engine remains operating in auto-ignition.

To avoid these drawbacks, it is necessary to apply an appropriate strategy of the EGR temperature and mixing control. With this control strategy, the same operating point N°2 can be run with smooth pressure rise (see Figures 9 and 10 "Controlled AI"), ultra low NOx level, slightly better efficiency. HC emissions slightly increase but remain at an acceptable low level, while the COV of IMEP reaches the low value of 2.5%.

The next Figures 10 show the corresponding mass fraction burnt and heat release. It clearly explains that in the case of "Uncontrolled AI " the auto-ignition appears too early and is too fast generating NOx emissions. In the "Controlled AI" case, timing of the combustion is much better and heat release rate quite reasonnable.

These results show that the gasoline auto-ignition combustion process here developed is not an uncontrolled one. Several control parameters are possible. The valve event configuration is probably the major parameter. Besides, the trapped EGR charge strategy is another important control parameter.

In the next sub-section, we are going to see that the trapped air charge strategy can be another control parameter.

THE COMBINED TRAPPED AIR CHARGE STRATEGY TO CONTROL THE CAI TIMING AND HEAT RELEASE

In this sub-section, we study the Reference point N°3 (4 bar IMEP @ 2000 rpm) which as shown by Figure 7 is in the border line of both the NA-CAI range and the Boosted-CAI range. In the case of NA-CAI, the auto-ignition is still obtained with the valve event optimized for the Reference point N°1. The EGR charge control strategy described previously is already applied and allow to reach a NOx level of 0.5 g/kWh. Nevertheless, the NA-CAI results presented in

Figures 11 show rather high cylinder pressure rise which indicates that NOx could be further reduced with an appropriate solution to smooth and delay the combustion.

The same Figures 11 show that with a boosted air charge strategy it is effectively possible to smooth the cylinder pressure rise and therefore drastically reduce the NOx emissions to a near zero NOx level.

The Figures 12 also show that similarly to the EGR charge strategy, the combined air charge strategy can be another parameter that can significantly affect the auto-ignition combustion timing and heat release rate.

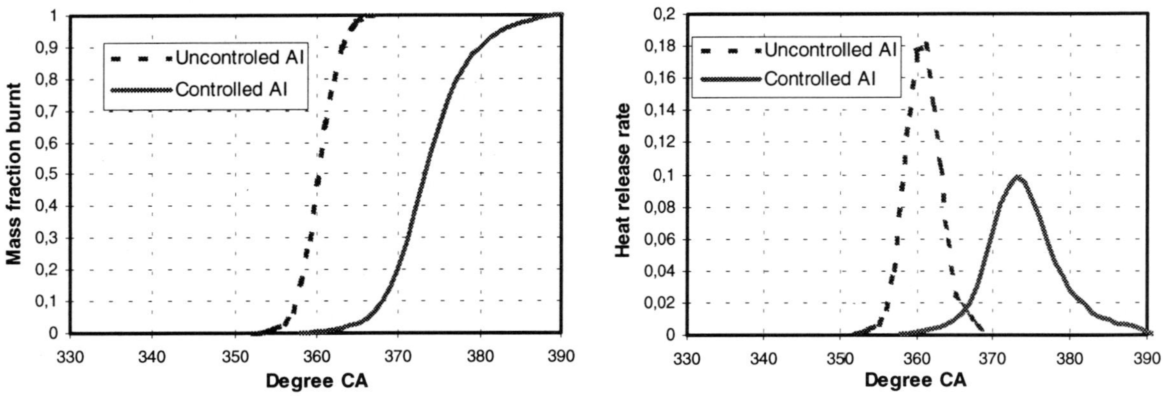

Fig. 10: Compared mass fraction burnt and heat release rate in two different conditions of auto-ignition control strategy (Reference point N°2: 4 bar IMEP @ 1500 rpm)

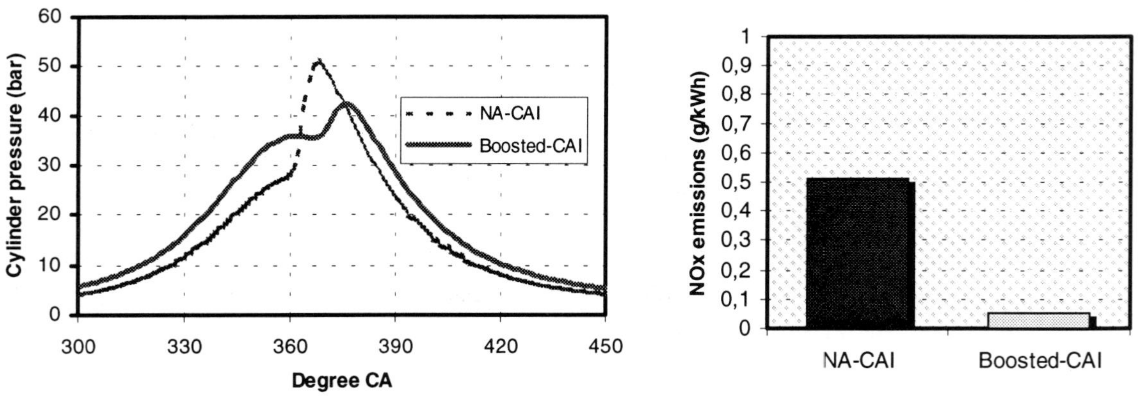

Fig. 11: Compared cylinder pressure traces and NOx emissions in CAI for two different conditions of intake air charging strategy (Reference point N°3: 4 bar IMEP @ 2000 rpm)

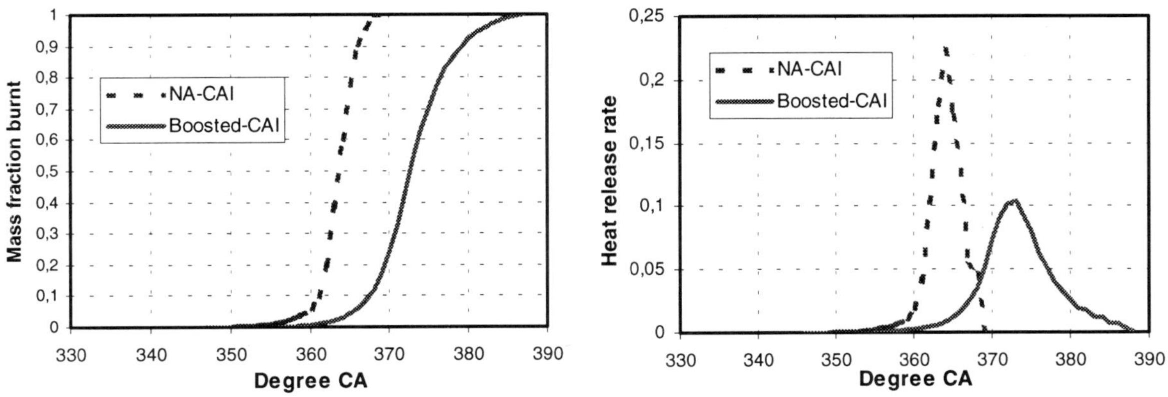

Fig. 12: Compared mass fraction burnt and heat release rate in CAI for two different conditions of intake air charging strategy (Reference point N°3: 4 bar IMEP @ 2000 rpm)

GASOLINE CAI COMBINED WITH DIRECT INJECTION: AN ALTERNATIVE TO THE STRATIFIED CHARGE DI?

All the results previously presented were obtained with Port Fuel Injection (PFI). Even if DI is not required for CAI, its combination with CAI can present some further advantages:

- It is easier to accept a compression ratio in the range of 11 to 12.5 with a DI engine from which we could expect a slightly wider CAI range compared to a more conventional CR of 10 to 11 in a PFI engine,
- DI timing can be another auto-ignition control parameter (by affecting the mixing between the EGR, the air and the fuel) as shown in the next Figure 13, which will slightly increase the flexibility of the EMS calibration.

The Figures 14 to 17 summarize typical preliminary results achieved in 4-cylinder gasoline engine with combined DI-CAI (in both lambda 1 and lean mixture operation) and compare those results with gasoline DISI (in both lambda 1 and stratified lean) for the same light load operating point which corresponds to urban driving conditions.

It can be seen in Figure 14, that compared to the gasoline DISI lambda 1 reference, the DISI stratified lean mode gives almost similar improvement in fuel consumption as the DI-CAI lean (with a slight advantage to the DI-CAI lean). The DI-CAI lambda 1 mode is a little bit above the DISI stratified lean and the DI-CAI lean, but still give about 15% improvement in fuel consumption.

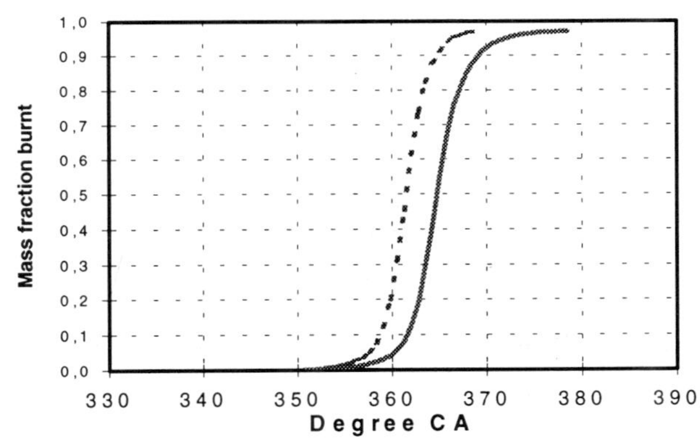

Fig. 13: Compared mass fraction burnt in CAI with Direct Injection for two different injection timing

The next Figure 15 shows the raw emissions of NOx and the drawbacks of the gasoline stratified lean DI. The high level of the gasoline DISI lambda 1 mode (as well as those already ultra low of the DI-CAI lambda 1) can be rather easily reduced to a level compatible with gasoline Euro 4 regulation by combining external EGR with three way catalytic aftertreatment.

It can also be seen that for this light load operating point, the DI-CAI lean combustion mode can meet the gasoline Euro 4 regulations without DeNOx aftertreatment. Its level of raw emissions of NOx is close to zero. It is not the case of the DI stratified lean combustion that will require a DeNOx aftertreatment of rather high efficiency to meet the gasoline Euro 4 level. Such aftertreatment would also necessitate the implementation of corresponding engine mana-gement strategies that could introduce some limited fuel penalties. This would modify the Figure 14 more in favour of DI-CAI lean compared to DISI stratified lean.

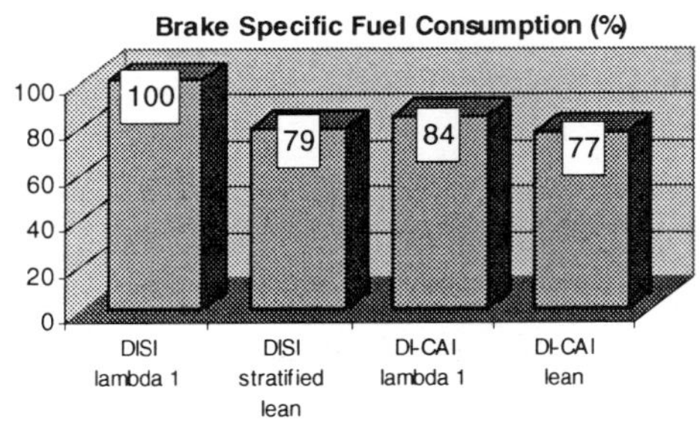

Fig.14: Compared fuel consumption between various gasoline DI combustion processes (2 bar BMEP @ 2000rpm)

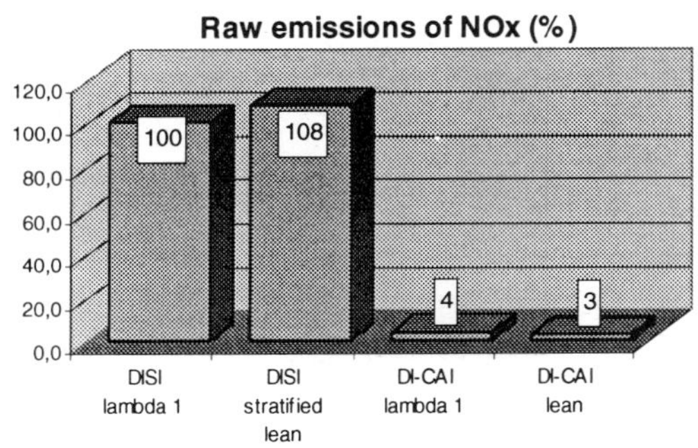

Fig. 15: Compared NOx emissions between various gasoline DI combustion processes (2 bar BMEP @ 2000 rpm)

What is generally identified as a drawback of HCCI combustion are the raw emissions of HC. It is generally true for gasoline HCCI engines when high compression ratio are used [11-13] and can be partly explained by the high air/fuel ratio used, the crevices and dead volumes generated by the high compression ratio combustion chamber shape. As it was already demonstrated in a previous paper in CAI combined with PFI [1], the Figure 16 confirms that one additional advantage of CAI is to achieve similar HC emissions than SI combustion, even slightly better with DI-CAI lean compared to the DISI lambda 1 reference. This interesting result can be obtained mainly thanks to the moderate and conventional compression ratio used in CAI. Besides, the HC emissions of the DISI stratified lean remain the highest.

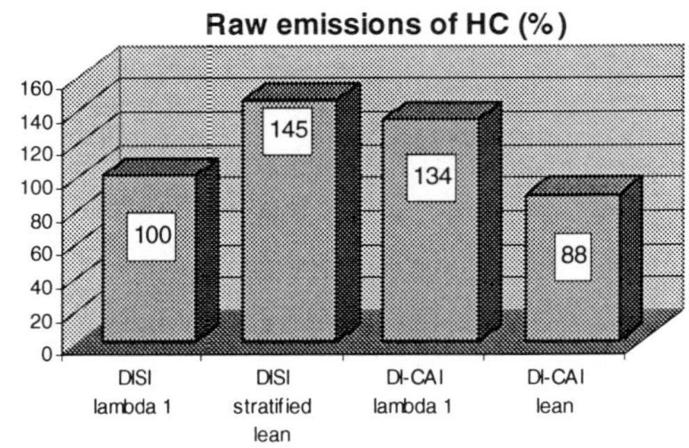

Fig. 16: Compared HC emissions between various gasoline DI combustion processes (2 bar BMEP @ 2000 rpm)

Regarding the CO emissions (Figure 17), they can be rather easily aftertreated by oxidation catalyst (or 3-way cat) technology in all the combustion modes. It can be seen that lean operation in DI-CAI helps to reduce further the CO emissions compared to lean operation in stratified DISI.

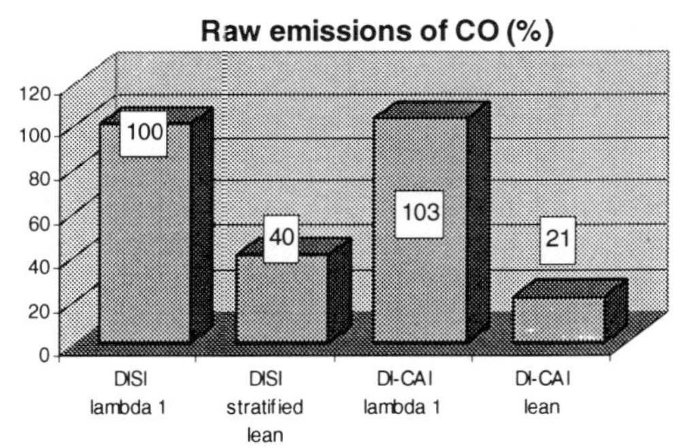

Fig. 17:Compared CO emissions between various gasoline DI combustion processes (2 bar BMEP @ 2000 rpm)

To conclude about this section, it clearly appears that lean operation in gasoline CAI combined with DI can be an alternative to the DI stratified lean combustion. It presents at least similar fuel economy advantages without the NOx drawback. In addition, without using a stratified air/fuel charging strategy it should be significantly less sensitive to carbon deposit which will have to be confirmed by further investigation work.

CONCLUSIONS

From the results presented in this paper, it has been possible to draw the following main conclusions:

- The research approach used to transfer the ultra low NOx 2-stroke engine gasoline controlled auto-ignition to the 4-stroke application was based on the combined and extensive use of the most advanced R&D tools such as 1D and 3D CFD in relation with optical diagnostics in operating engines,
- In spite of the complexity of the processes involved and thanks to the scientific and sophisticated methodology adopted, it has been possible to demonstrate the achievement of gasoline CAI™ (Controlled Auto-Ignition) in automotive 4-stroke engine design. It opens new scope for clean and efficient energy conversion by powertrains, as shown by the potential of "one digit ppm" of raw emissions of NOx combined with the 15 to 20% fuel economy improvement achieved in some operating conditions,
- In spite of inoperating combustion control by spark timing, this new CAI combustion process can be controlled (in terms of combustion timing and heat release rate) by several other parameters. Among them, the valve events configuration is the most important one, the other important control parameters being the EGR charge strategy, the air charge strategy, the Direct Injection timing (when CAI is combined with DI),...
- Some preliminary lean operation results obtained in 4-cylinder gasoline engine combining CAI with DI show that this combination could represent an alternative way to the DISI stratified lean combustion. It presents robust ultra low NOx emissions with at least similar fuel economy advantages. It would nevertheless require further investigation work to assess other potential advantages and drawbacks compared to the DISI stratified lean combustion,
- By allowing to run a gasoline engine in lean burn mode without requiring any fuel-sensitive DeNOx aftertreatment technology, the CAI combustion process can further enhance the already known Variable Valve Actuation (VVA) technologies benefits in terms of fuel efficiency and low CO2 without needing any other engine modification or redesign. It is therefore expected that the coming new generation of VVA technologies and associated engine management system, will make possible the introduction of the CAI combustion process (in at least some operating range) in the automotive gasoline engines of the short to medium term future.

DEFINITIONS, ACRONYMS, ABBREVIATIONS

4-SPACE: 4-Stroke Powered gasoline Auto-ignition Controlled combustion Engine
BMEP: Brake Mean Effective Pressure
CAI: Controlled Auto-Ignition
COV of IMEP: Standard deviation of IMEP divided by IMEP (in percentage)
CR: Compression Ratio
DI: Direct Injection
DISI: Direct Injection Spark Ignition
EMS: Engine Management System
HCCI: Homogeneous Charge Compression Ignition
IMEP: Indicated Mean Effective Pressure
NA: Naturally Aspirated
PFI: Port Fuel Injection
SI: Spark Ignition
VVA: Variable Valve Actuation

ACKNOWLEDGEMENTS

The European Commission is deeply acknowledged by the authors for its funding of the "4-SPACE" project.
The authors also acknowledge all the IFP teams which contributed in obtaining the information and data used to prepare this paper, with especially the teams involved in the 3D CFD calculations of CAI engine configurations, in the CAI optical engine diagnostics, in the gasoline CAI combustion process development and in its combination with DI technology (Gaetan Monnier and Alain Ranini).

REFERENCES

[1] DURET P. and DOUAUD A. - New Combustion Processes : Homogeneous Diesel and Auto-Ignition Gasoline Towards Zero Emission with Major Engine and Fuel Technology Challenges GPC'2001 Detroit - 5-7 June 2001

[2] Automotive Engineering – October 1999

[3] DURET P. and VENTURI S. - Automotive Calibration of the IAPAC Fluid Dynamically Controlled Two-Stroke Combustion Process

SAE 960363 - Detroit

[4] PETIT A., LAVY J., MONNIER G. and MONTAGNE X. - Speciated Hydrocarbon Analysis: A Helpful Tool for Two-Stroke Engine Development - IFP International Seminar; Paris 1993.

[5] LAVY J., DURET P. - IFP International Consortium Project on the Effects of Fuels and Additives on Controlled Auto-Ignition - IFP Consortium reports 1995-97

[6] LAVY J.; DABADIE JC.; ANGELBERGER C.; WILLAND J.; JURETZKA A.; SCHAFLEIN J.; MA T.; LENDRESSE Y.; SATRE A.; SCHULZ C.; KRAMER H.; ZHAO H.; DAMIANO L. - Innovative Ultra-low NOx Controlled Auto-Ignition Combustion Process for Gasoline Engines: the 4-SPACE ProjectSAE 2000-01-1837 Paris, 2000

[7] AUST R. - Powertrain Research and Development: Needs, Trends and Visions - EUCAR Conference - Brussels, November 22, 2000.

[8] DURET P.; LAVY J.- Gasoline Controlled Auto-Ignition (CAI™) : potential and prospects for future automotive application - ImechE Conference: 21st Century Emissions Technologies

London, 4 December 2000

[9] DURET P,LAVY J.; SCHAFLEIN J.; MA T.; GIMBRES D.; ZHAO H.; SCHULZ C - 4-SPACE Synthesis Report European Commission, 2001

[10] LI J., ZHAO H., LADOMMATOS N. and MA T.- Research and Development of Controlled Auto-Ignition (CAI) Combustion in a 4-stroke Multi-cylinder Gasoline Engine - SAE paper 01FL-393, Detroit, March 2001.

[11] AOYAMA T., HATTORI Y., MIZUTA J., SATO Y. - An Experimental Study on Premixed-Charge Compression Ignition Gasoline Engine

SAE 960081

[12] STOCKINGER, H. SCHÄPPERTÖNS, P. KUHLMANN - Versuche an einem gemischansaugenden Verbrennungsmotor mit Selbstzündung (Investigations on a gasoline Engine working with Self-Ignition by Compression)

MTZ Motortechnische Zeitschrift 53 (1992) 2 (p.80-85)

[13] J. WILLAND, G. VENT, F. WIRBELEIT, G. WENNINGER, H.P. GULDE - HCCI process – the future for automotive engines? - International Symposium of Diesel Engine Technology for Passenger Cars Porto Cervo, Italy – October 12-13, 2000

[14] GRAF N., GRONKI J., SCHULTZ C., BARITAUD T., CHEREL J., DURET P., LAVY J. - In-Cylinder Combustion Visualization in an Auto-Igniting Gasoline Engine using Fuel Tracer and Formaldehyde LIF Imaging - SAE paper 2001-01-1924

[15] LI J.; ZHAO H.; LADOMMATOS N.; MA T. - Experimental Investigation of Controlled Auto-Ignition (CAI) Combustion in a 4-Stroke Multi-Cylinder Gasoline Engine and Drive Cycle Simulations - to be published by Editions Technip - IFP International Congress, November 26-27, 2001

[16] DURET P. and LAVY J.

Process for controlling self-ignition in a 4-stroke engine - US Patent 6 082 342

[17] DABADIE J.C. and LAVY J. - Controlled self-ignition combustion process and associated 4-stroke combustion engine with transfer conduit between cylinders and valve for it. - Patent Application EP 0953 745

A New Generation of Engine Combustion Processes for the Future?
© P. Duret (Editor) and Editions Technip, Paris, 2001, pp. 115 - 124
27 rue Ginoux, 75015 Paris

EXPERIMENTAL INVESTIGATION OF CONTROLLED AUTO-IGNITION (CAI) COMBUSTION IN A 4-STROKE MULTI-CYLINDER GASOLINE ENGINE AND DRIVE CYCLE SIMULATIONS

Tom Ma[1] , Hua Zhao[2], Jian Li[2], Nicos Ladommatos[2]
1 Ford Motor Company 2 Brunel University

ABSTRACT

Controlled Auto-Ignition (CAI) combustion has been realised in a production type 4-stroke multi-cylinder gasoline engine. The engine is based on a Ford 1.7l Zetec-SE 16v engine using substantially standard components modified only in design dimensions in order to significantly increase the trapped residuals. The engine is also equipped with Variable Cam Timing (VCT) on both the intake and exhaust camshafts. CAI combustion is achieved by only using hot internal residuals, and a range of loads between 0.5 and 4 bar bmep and engine speeds between 1000 and 3500 rpm are mapped for CAI fuel consumption and exhaust emissions. The measured CAI results are compared with those of spark ignition (SI) combustion in the same engine at the same speeds and loads. The comparison shows that a more than 30% reduction in bsfc and up to 99% reduction in NOx at low loads can be reached. European New Emission Drive Cycle (NEDC) is simulated for a vehicle equipped with such a CAI combustion engine. Fuel consumption and emissions are compared with an SI engine vehicle undergoing the same drive cycle.

INTRODUCTION

Controlled Auto-Ignition (CAI) combustion, which is different from both gasoline SI and diesel CI combustion, is a new combustion process for internal combustion engines. It has low cyclic variations [1], better fuel economy, and most important of all much less NOx emissions [2,3,4] compared with SI engines.

The investigation of CAI combustion was started in 2-stroke engines [5]. Significant progress has been made in the studies and application of CAI combustion in 2-stroke engines in which CAI is achieved by using a large mount of hot residuals to raise the charge temperature [6, 7]. Due to the gas exchange characteristics, conventional 4-stroke engines are very difficult to trap the required amount of hot residual gas in the cylinders, hence other methods such as intake air heating [8] and increasing compression ratio [9] etc. were used. However, these methods by researchers in previous studies are extremely difficult to be applied to a practical gasoline engine. Recently, special camshaft configurations [3,4] and fully variable valve train [10] are employed in 4-stroke engines in which a large amount of hot residual gas is successfully trapped in the cylinders in order to achieve CAI combustion.

In the current work, special camshafts are used in order to retain a large amount of hot residuals in the cylinders in a 4-stroke multi-cylinder gasoline engine. The significantly increased trapped residuals alone are sufficient to achieve CAI combustion in this engine. In addition, Variable Cam Timing (VCT) systems are used. CAI combustion is obtained in a range of loads between 0.5 and 4 bar BMEP and engine speeds between 1000 and 3500 rpm. Fuel consumption and engine-out emissions are measured and compared with those of spark ignition (SI) combustion in the same engine. European New

Emission Drive Cycle (NEDC) simulations are conducted assuming that the engine is able to run CAI combustion in the CAI range and SI combustion outside the CAI range without any disruptions. In the simulation the cumulative fuel consumption and emissions after treatment in the whole cycle are obtained for a Ford Mondeo vehicle equipped with the CAI/SI engine. The simulated data are compared with the figures from a standard Mondeo 1.8 undergoing the same cycle.

ENGINE SETUP

Table 1 Engine specifications.

Engine type	Bore (mm)	Stroke (mm)	Capacity (cm^3)	Compression ratio	Fuel
Ford 1.7 Zetec-SE	80	83.5	1679	10.3	95 RON

The engine used in this work is a Ford 1.7L Zetec-SE 16V four-cylinder, multi-point port injection gasoline engine. Its specifications are listed in Table 1. In addition to the standard VCT system installed on the intake side, another VCT system is cobbled onto the exhaust side to give the ability to adjust exhaust cam timing when the engine is running. The duty cycle of the solenoid valves in the VCT system is controlled by a 10 Hz pulse generator. Another major modification to this engine is that special camshafts are used in order to significantly increase the trapped residuals [11]. A Webcon Alpha-Plus ECU system, whose strategies are based on engine speed and throttle angle, is used to control the fuelling and ignition. Since all the tests are running at wide open throttle conditions, a dummy throttle is used and the signal from the dummy potentiometer is fed into the ECU to obtain the required fuelling. Also in this system there are facilities to finely adjust the fuelling and ignition timing manually. A sequential fuelling facility in this system is essential to provide the same fuelling events, hence same fuel preparation, for each of the four cylinders. An UEGO sensor is used to monitor the air fuel ratio and fine fuelling adjustment is made manually if necessary to maintain the required lambda value.

An eddy-current dynamometer is employed to set the engine running at a desired speed, and a load cell is used to obtain the output torque. NOx and unburnt hydrocarbons are measured by a SIGNAL heated vacuum NOx analyser and a SIGNAL FID total hydrocarbon analyser respectively. CO and CO2 are measured by an OLIVER K650 MOT analyser.

A Kistler Type 6125A pressure transducer is installed in one of the four cylinders to measure the in-cylinder pressure. The signal from this pressure transducer is fed into a PC based data acquisition and heat-release analysis system developed at Brunel. The program generates an array of pressure traces against crank angle, and from the pressure records it calculates the heat release data, indicated mean effective pressure (IMEP) and coefficient of variation of IMEP. Four thermocouples are installed in each of the four exhaust ports to monitor the exhaust temperatures and detect misfire should it happen.

EXPERIMENTAL PROCEDURE

The engine was started in the spark ignition mode. After one or two minutes the engine is moderately warm and CAI combustion appears, but it is unstable, i.e. CAI and SI combustion occur intermittently for several cycles. Misfire is observed occasionally during this period. A few minutes later when the coolant temperature reaches about 50 – 60°C CAI combustion stabilises. However, to minimise the effect of coolant temperature on CAI combustion, all the tests are conducted when the coolant temperature is 90°C or over.

There are two ways to determine if CAI combustion is actually happening in the cylinders. One is to disconnect the ignition coil hence there is no spark in the cylinders. If the ignition coil is disconnected and the engine keeps running it proves that the combustion in the cylinders is CAI. The other way is to check the in-cylinder pressure trace. Pressure records of CAI combustion, which usually display a distinctively sharp rise and much higher magnitude near TDC, are different from those of SI combustion (Figure 1). The latter way is usually used as the spark is kept on through out the tests, and in the case that a CAI misfire happens the spark could ignite the mixture in the subsequent cycles. Also keeping the spark on in CAI mode does not cause any problems. Actually the spark has little effect on CAI combustion, changing the spark timing substantially shows virtually no effects on combustion. And the CAI engine tests are carried out at wide-open throttle.

EXPERIMENTAL RESULTS OF CAI COMBUSTION

Results are obtained when CAI combustion has stabilised and the coolant temperature has reached 90°C or over. The air fuel ratio is kept at lambda 1 and the engine load is varied by using different cam timings. Figure 2 shows the speed and load range within which CAI combustion is achieved in the engine. At each speed there is a minimum BMEP, which is limited by misfire. At these conditions the load is very low and there is too much residual gas. As a result, the exhaust temperature is so low that the mixture fails to be compression ignited. The charge could not be spark ignited either, as too much residual gas is in the mixture. At higher speeds, the lower limit is shifted to lower load, maybe due to less heat losses at higher speeds. There is also a maximum BMEP at each speed as a consequence of restrictions of gas exchange by the special cams. At these conditions the air flow is the maximum the engine can breathe in. As speed increases the maximum load reduces due to the fact that at higher speed, there is less time for gas exchange process hence more residuals are trapped in the cylinders and less fresh charge is inducted. The CAI window becomes smaller as engine speed increases. At 3500 rpm the CAI range is only from 0.4 to 1.1 bar BMEP. CAI range for speed higher than 3500 rpm is even narrower and not included.

Figure 3 shows the brake specific fuel consumption across the CAI range. BSFC mainly depends on load and does not change very much with engine speed. For the same speed BSFC reduces as the load increases. But for the same load at different speeds BSFC shows little change, especially at low load conditions.

Figures 4 – 6 give the NOx, CO and HC emissions of CAI combustion in the speed and load range. For a given speed, as load increases NOx increases, CO and HC decrease as expected. The variations of each of the three emission components with speeds are different. NOx does not change very much with speed at low and high speeds, but increases with speed in the mid-speed range. Speed hardly shows an effect on CO, especially at high speeds. HC decreases as speed increases suggesting that combustion at high speeds is better and more complete. The reasons why the three emission components behave like this in the speed and load range are not very clear and detailed combustion features of CAI need to be investigated.

The above CAI results are compared with those of SI combustion in the same engine at the same speeds and loads. Figure 7 shows the percentage change of CAI BSFC relative to SI BSFC. In the whole CAI speed and load range, there is a BSFC improvement for CAI. At higher loads, the fuel economy improvement is small, only about 5 – 10% at 2.5 bar BMEP. As the load decreases the improvement becomes significant and it reaches about 30% at 1 bar BMEP. Further fuel consumption reduction is achieved at lower loads. The better fuel consumption comes from two sources. The first is that the engine is unthrottled with CAI combustion so there is a reduction in pumping work. This effect is much more important at low loads because at these conditions SI engines are usually highly throttled, the reduction in pumping work is enormous for

Table 2 NEDC cruise point specifications.

Vehicle Speed (km/h)	0	15	32	35	50		70	100	120
Engine speed (rpm)	880	1742	2324	1723	2461	1747	1826	2608	3130
BMEP (bar)	0	0.170	0.378	0.592	0.792	1.115	2.136	3.397	4.437
In CAI range	no	no	no	no	yes	yes	yes	no	no
Duration %	23.7	4.2	8.1	5.3	4.1	6.0	8.5	2.5	0.8

unthrottled CAI combustion. At higher loads the throttling of SI engines is reduced and less improvement in BSFC is observed. The other reason for BSFC improvement is that during CAI combustion the sharp pressure rise occurs near TDC and it renders CAI combustion like constant-volume combustion, which is the most efficient heat addition in an Otto cycle process. In figure 1 the typical P-V diagram of the power cycles of CAI and SI combustion is shown and the constant-volume like feature can be clearly seen on the CAI curve.

Figure 8 shows the percentage improvement in NOx emissions. In the CAI range 90 – 99% reduction in NOx emissions are observed. That means NOx emissions from CAI are only one hundredth to one tenth of those from SI combustion. This immense reduction is similar to the results of an effective catalyst. It may be argued that after treatment may not be necessary in this range for CAI combustion. The huge reduction in NOx emissions is a result of much reduced combustion temperatures. Due to the large amount of residual gas (especially at low loads) in the cylinders the typical temperatures of CAI combustion are around 1800 K while typical SI combustion temperatures are around 3000 K. In the much cooler CAI combustion there is hardly any NOx formed. This is true at low load conditions during which the readings from NOx analyser are close to figures of ambient air. At higher loads more energy is released during combustion and the temperature of the total mixture is higher, hence NOx is higher. But even at the upper end of the load the NOx from CAI is only about 10% of NOx from SI. It can be seen from figure 9 that CAI combustion reduces CO emissions by 10 to 40% depending on speed and load. At low speed and high load conditions CO reduction is higher. However, HC emissions from CAI combustion are worse than in SI combustion. In figure 10, a 50% to 160% increase in HC is found for CAI combustion. That means HC emissions from CAI are 1.5 to 2.6 times of those from SI combustion. The reason for the HC increase is not absolutely clear. It might be due to the low CAI combustion temperatures. At lower loads the combustion temperatures are even lower and more HC increases are recorded. However the actual cause of the HC emission increase needs to be further investigated

DRIVE CYCLE SIMULATIONS

The results above strongly suggest the possibility of a practical engine for use in a road vehicle, in which the gas exchange process could be switchable to produce either CAI combustion or SI combustion depending on the drive demand. A vehicle simulation program is used to model a medium size vehicle, a new Ford Mondeo, with a hypothetical CAI/SI switchable engine driven through the European New Emission Drive Cycle (NEDC) using measured data from the above CAI engine combined with known data from the same production SI engine equipped only with a three-way catalyst. NEDC, which starts from cold and lasts for 1180 seconds, consists of four urban phases and one high way phase (figure 11). It covers a distance of about 11 km. The coolant and oil temperatures of the engine are at 20°C at the beginning of the cycle. They gradually increase and stabilise. Figure 12 shows the engine speed and load for the standard SI engine in a Mondeo undergoing the cycle. There are eight cruise points corresponding to the idle condition and seven cruise speeds (table 2). Other conditions represented by crosses in figure 12 are transient conditions. In the CAI range when the engine is fully warmed up there are three cruise points which represent cruise speeds of 50 and 70 km/h respectively and are about 18.6% of the total cycle time. There are also some transient points in the CAI range.

To conduct the simulation of the vehicle driven over NEDC, the following assumptions are made. (1) The engine is the same as the standard SI engine and is switchable between SI and CAI without any disruptions and response delays, i.e. within the CAI range the engine runs CAI combustion and outside the range the engine runs SI combustion. The switch between SI and CAI can be as frequent as necessary. (2) The CAI/SI engine has the same rate of engine warm-up and same cold engine friction as the standard SI engine. Cold engine can not use CAI below 60°C, even the speed and load is within the CAI range. Only when the coolant temperature is 60°C or over can the engine run CAI combustion in the CAI range. (3) The vehicle with the CAI/SI engine is the same vehicle as one with a standard SI engine. It has the same gearbox, same final drive ratio and same exhaust after-treatment system, which is only a three-way catalyst. The catalyst lights off in 20 seconds after start and its fully warmed-up efficiency is 98%.

Figure 13 shows the CAI usage during the cycle according to the CAI range and above assumptions. CAI combustion starts at 264th second when the coolant temperature reaches 60°C. When the engine is warming up, the BMEP at the cruise points are higher than the values in figure 12, due to higher friction at cold conditions. Therefore, at cruise speeds of 32 and 35 km/h which are just below the CAI boundary, the engine runs into CAI range. In the final urban phase when the engine is fully warmed-up, the BMEP at cruise speeds 32 and 35 km/h decreases to the values in figure 12, which are below the CAI boundary, so the engine runs SI combustion.

Figures 14 to 17 show the cumulative fuel consumption and after-catalyst emissions over NEDC. The total fuel consumption over NEDC for the SI/CAI is 598 grams, which is about 4.7% less than 627 grams, the standard engine consumption.

Table 3 Percentage change of CAI/SI relative to SI over NEDC.

Drive modes	Change in fuel consumption %	Change in NOx %	Change in CO %	Change in HC %
NEDC complete cycle	-4.7	-12.7	-3.7	16.9
Urban phase	-4.6	-9.7	-2.0	10.7
Highway phase	-5.0	-16.6	-8.0	40.8
50 km/h cruises	-17.9	-67.7	-16.9	109.2
70 km/h cruises	-11.1	-91.5	-20.0	88.2

After the engine coolant reaches 60°C at which the engine is able to run CAI mode, there is a reduction in NOx for the CAI/SI engine. This reduction becomes greater at highway phase. And over the whole NEDC the NOx reduction is 12.7%.

Table 3 lists the percentage change of the CAI/SI engine relative to the SI engine in fuel consumption and after-catalyst emissions, where a negative figure represents a reduction and a positive value indicates an increase. It is not surprising to see that in highway phase, of which CAI mode consists a large portion, the reductions in fuel consumption and NOx and CO emissions are higher. However, there is a large increase in HC emissions in highway phase. From figure 13 it can be seen that during the entire 70 km/h cruise period the engine runs CAI. The changes in the fuel consumption and emissions are significant, especially the 11.1% reduction in fuel consumption and 91.5% decrease in NOx are very impressive. The first 50 km/h cruise cycle is at cold conditions and the engine runs SI combustion. Therefore the reduction in NOx over all of the 50 km/h cycles is only 67.7%. However, the fuel consumption decrease is as high as 17.9%, thanks to the large reduction in BSFC at low load conditions at this cruise speed (figure 7 and 12). The reduction in NOx in the complete NEDC is 12.7%, may be lower than one expects. This is because that the highest speed and load part of NEDC (100 and 120 km/h cruises), during which NOx emissions are very high, are not in the CAI range hence there is no NOx reductions in this important period. As to HC emissions, higher increase for CAI combustion appears at low load conditions, which are exactly the CAI usage conditions in the vehicle simulation. Therefore overall HC emission increase is as high as 16.9%. Even so, the average after-catalyst NOx emissions for the CAI/SI vehicle over NEDC are only 0.022 g/km, far less than the EURO 4 legislation figure of 0.1 g/km.

Although the improvement in fuel economy and NOx emissions under CAI mode is promising, the overall reductions in the complete NEDC are much less due to the fact that it is unable to use CAI combustion in a large portion of the NEDC. It is important in the future to enlarge the CAI range and hopefully bring the majority part of the drive cycle into the CAI boundary.

CONCLUSIONS

1. CAI combustion is achieved in a 4-stroke multi-cylinder gasoline engine with substantially standard components. The CAI range is between 1000 and 3500 rpm with the load varying between 0.5 bar and 4 bar BMEP.
2. Within the CAI range, the reduction in BSFC varies from 5% to more than 30%, depending upon speed and load, while NOx reductions are between 90% and 99%. CO emissions from CAI combustion are slightly lower than those from SI combustion. But unburned hydrocarbons from CAI are 0.5 to 1.6 times higher than those from SI.
3. Simulations of a vehicle with a CAI/SI engine over NEDC have been carried out. The results indicate about 4.7% reduction in fuel consumption and 12.7% decrease in after-catalyst NOx emissions over the entire drive cycle.

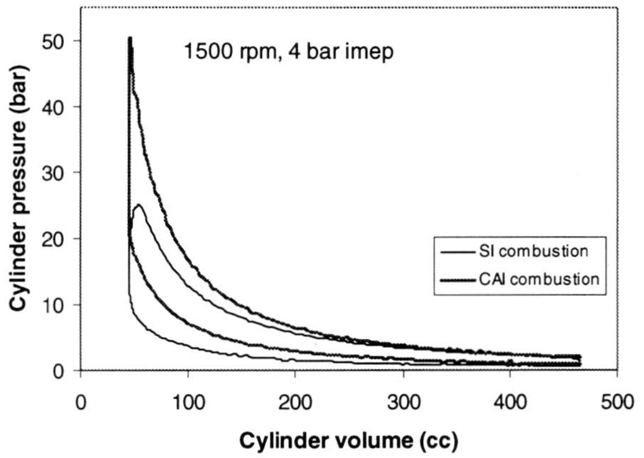

Figure 1 Typical pressure traces of SI and CAI combustion

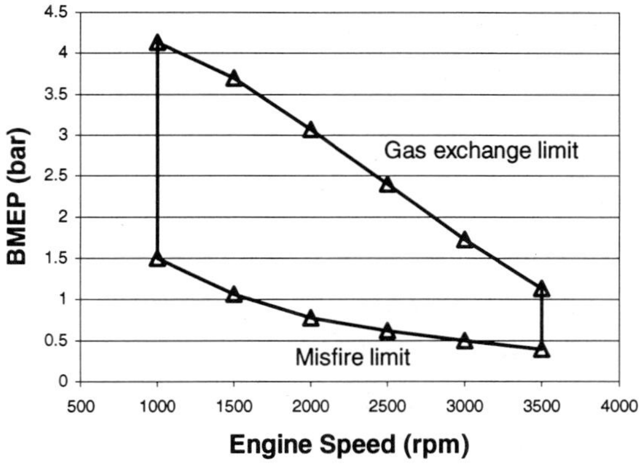

Figure 2 CAI speed and load range

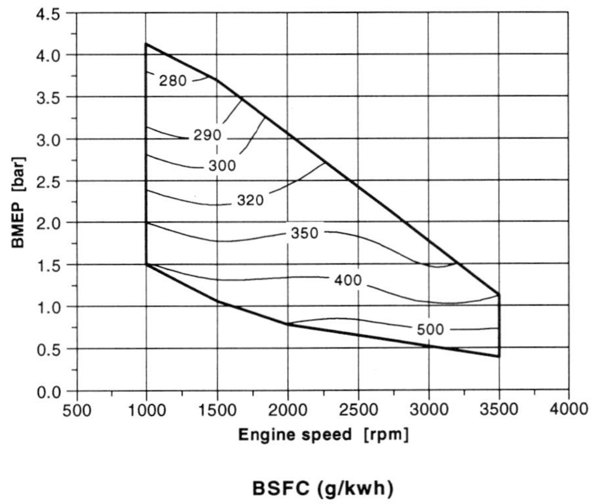

Figure 3 Fuel consumption of CAI

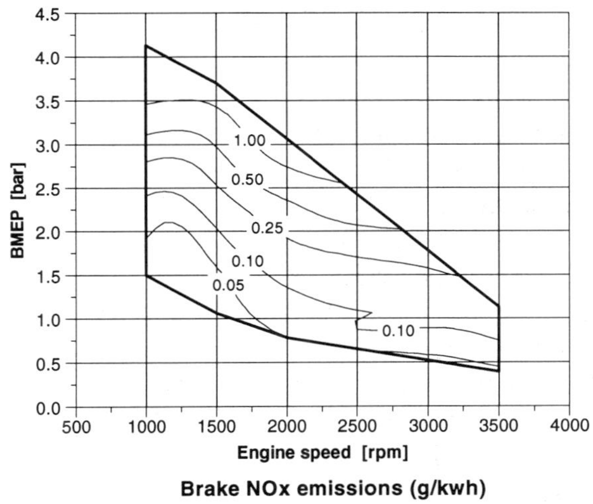

Figure 4 Brake NOx emissions of CAI

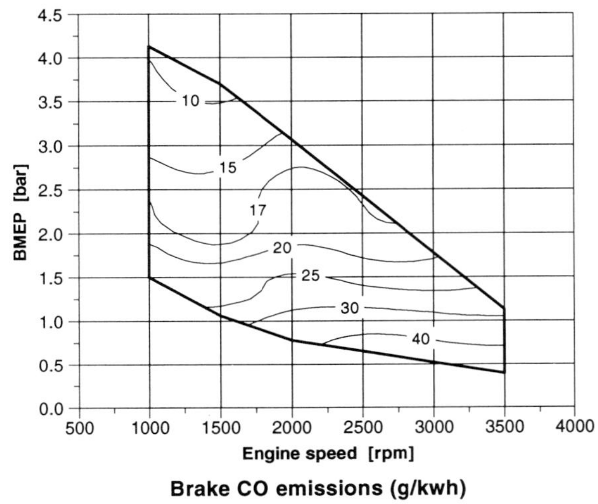

Figure 5 Brake CO emissions of CAI

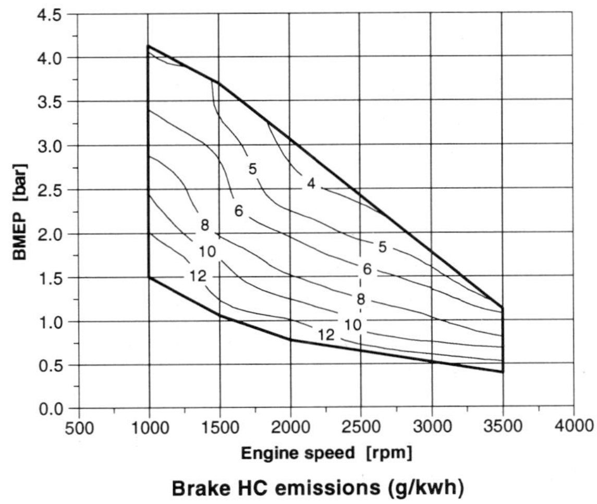

Figure 6 Brake HC emissions of CAI

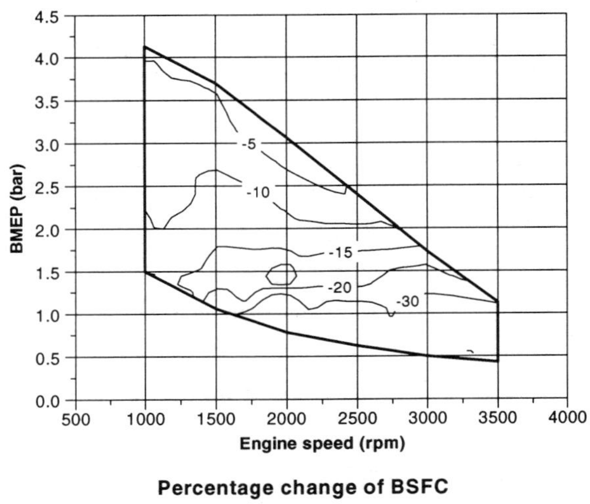

Figure 7 Percentage change of BSFC

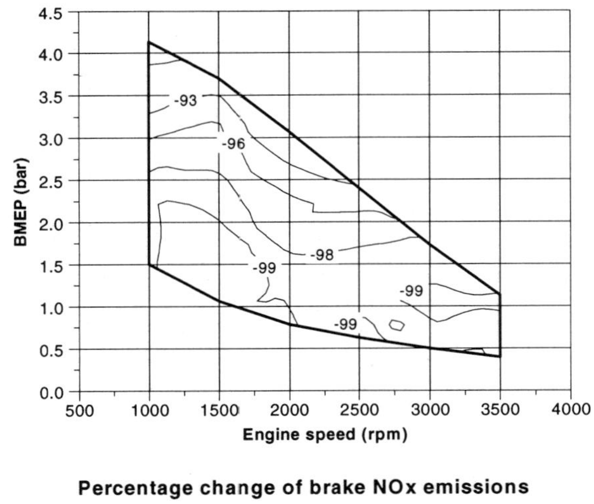

Figure 8 Percentage change of brake NOx emissions

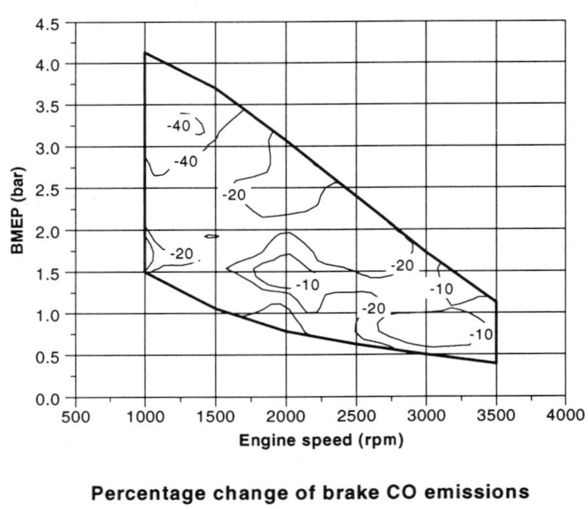

Figure 9 Percentage change of brake CO emissions

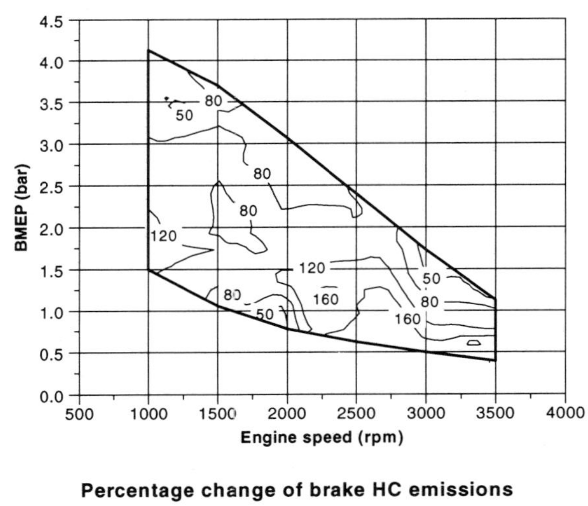

Figure 10 Percentage change of brake HC emissions

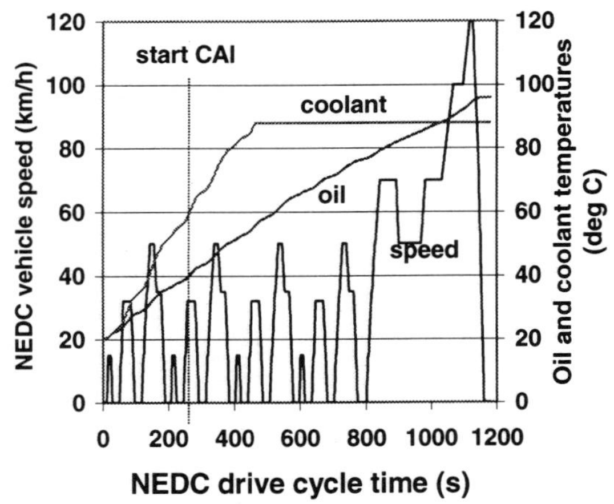

Figure 11 NEDC vehicle speed and coolant and oil temperatures

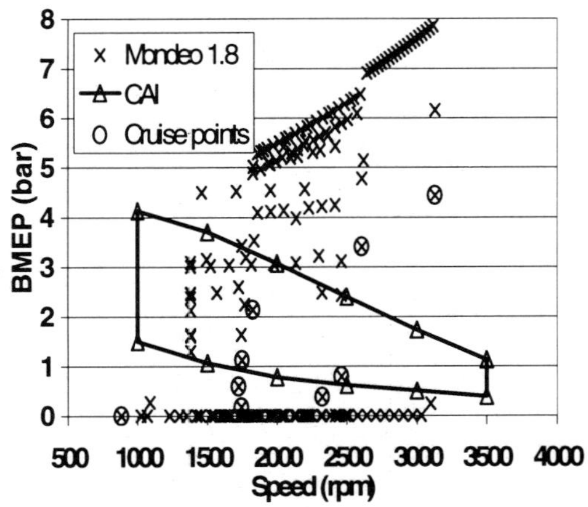

Figure 12 CAI window and NEDC speed/load range

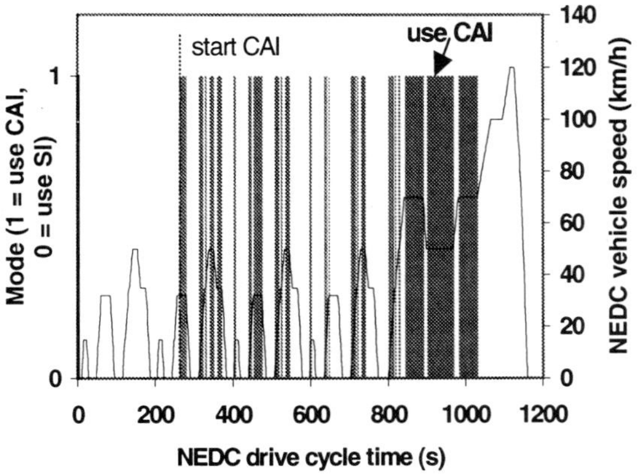

Figure 13 Switching strategies for SI+CAI engine

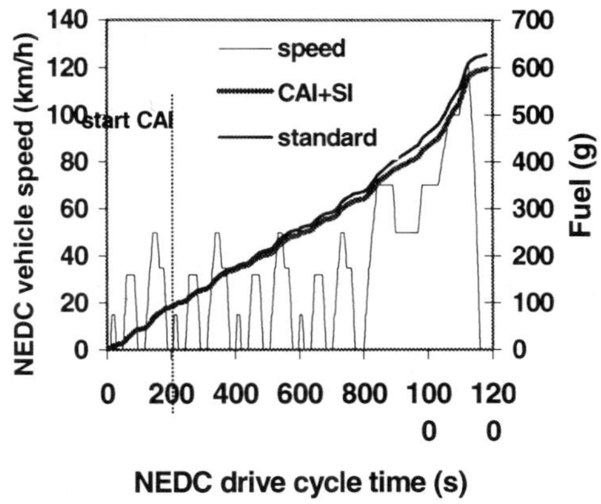

Figure 14 Cumulative fuel consumption of Mondeo vehicle driven over NEDC

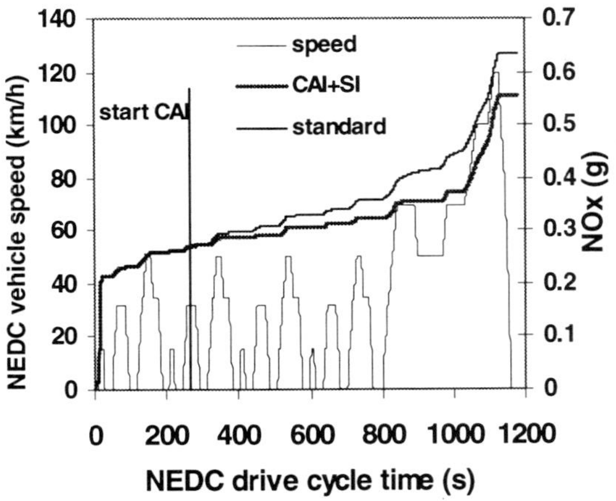

Figure 15 Cumulative NOx after-catalyst of Mondeo vehicle driven over NEDC

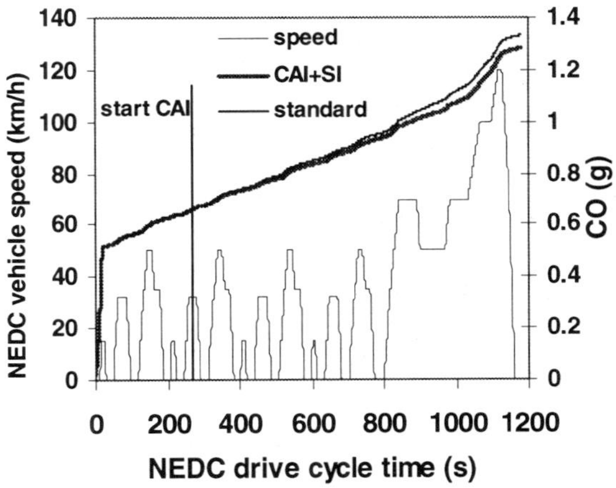

Figure 16 Cumulative CO after-catalyst of Mondeo vehicle driven over NEDC

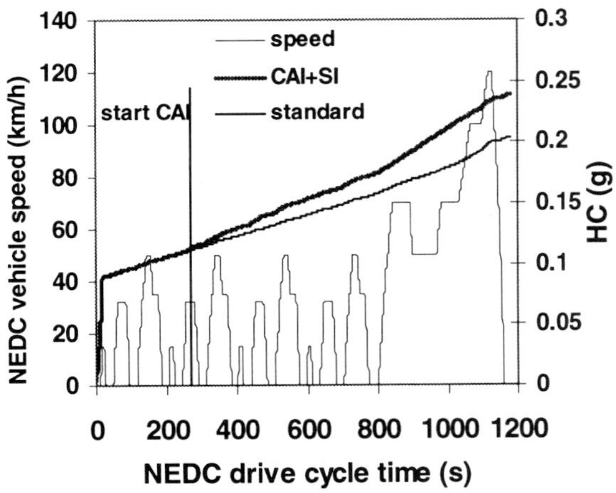

Figure 17 Cumulative HC after-catalyst of Mondeo vehicle driven over NEDC

ACKNOWLEDGEMENTS

This work is a part of the 4-SPACE project, and the authors would like to thank the European Commissions for its financial support to the 4-SPACE Consortium, and other partners in this Consortium for their collaboration.

REFERENCES

1. Iida, N., Ichikura, T., Kase, L. and Yoshiteru, E., Self-ignition and combustion stability in a methanol fuelled low heat rejection ceramic ATAC engine - analysis of cyclic variation at high wall temperatures and lean burn operation, JSAE Review 18, pp233-240, 1997
2. Oakley, A., Zhao, H., Ladommatos, N. and Ma, T, Experimental studies on controlled auto-ignition (CAI) combustion in a 4-stroke gasoline engine, international conference on 21st century emissions technology, C588/019/2000, IMechE, 2001
3. Lavy, W., Dabadie, J., Angelberger, C., Duret, P.(IFP), Willand,J., Juretzka, A., Schaflein,J.(Daimler-Chrysler), Ma, T. (Ford), Lendresse, Y., Satre, A. (PSA Peugeot Citroen), Schulz, C., Kramer, H. (PCI – Heidelberg University), Zhao, H., Damiano, L. (Brunel University), Innovative ultra-low NOx controlled auto-ignition combustion process for gasoline engines: the 4-space project, SAE paper 2000-01-1837, 2000
4. Li, J., Zhao, H., Ladommatos, N. and Ma, T, Research and development of Controlled Auto-Ignition (CAI) combustion in a 4-stroke multi-cylinder gasoline engine, SAE paper 01FL-393, 2001
5. Noguchi, M., Tanaka, Y., Tanaka, T. and Takeuchi, Y., A study on gasoline engine combustion by observation of intermediate reactive products during combustion, SAE paper 790840, 1979.
6. Ishibashi, Y and Masahiko, A., Improving the exhaust emissions of 2-stroke engines by applying the activated radical combustion, SAE paper 960742, 1996
7. Duret, P. and Venturi, S., Automotive calibration of the IAPAC fluid dynamically controlled 2-stroke combustion process, SAE paper 960363
8. Thring, R., Homogeneous charge compression ignition (HCCI) engines, SAE paper 892068, 1989
9. Christensen, M. and Hultqvist, A., Demonstrating the multi fuel capability of a homogeneous charge compression ignition engine with variable compression ratio, SAE paper 1999-01-3679, 1999
10. Law, D., Kemp, D., Allen, J., Kirkpatrick, G. and Copland, T., Controlled combustion in an IC engine with a fully variable valve train, SAE paper 2000-01-0251, 2001
11. Duret, P. and Lavy, J., Process for controlling self-ignition in a 4-stroke engine, US patent 6 082 342

ACTIVATED RADICALS COMBUSTION PROCESSES

A New Generation of Engine Combustion Processes for the Future?
© P. Duret (Editor) and Editions Technip, Paris, 2001, pp. 127 - 140
27 rue Ginoux, 75015 Paris

RADICAL IGNITION IN LOW COMPRESSION RATIO D.I. ENGINES ENABLED BY SONEX PISTON DESIGN

David A. Blank, Andrew A. Pouring
Sonex Research, Inc.

ABSTRACT

Using a variety of fuels, this study numerically and experimentally examines the nature and potential of the low and high temperature mechanisms responsible for homogeneous combustion in I.C. engines using intermediate and radical chemical species produced in micro-chambers of Sonex Combustion System (SCS) pistons. The study broadens the distinction between the currently popular thermally induced variant of homogeneous combustion: HCCI; and the emerging radical induced variant of homogeneous combustion: SCRI (Stratified Charge Radical Ignition), with the later being the focus of the current work. For the numerical portion of the study, a detailed chemical kinetics approach is adopted using CHEMKIN-II and integrated (augmented) with turbulent combustion KIVA-II flow-field calculations. Results point conclusively to the phenomenological importance of the dominant effects of several intermediate and radical species formed and retained in frozen equilibrium (chemically inert at lower pressures and temperatures) in the SCS micro-chamber. These dominant species are shown to enhance ignition by changing the oxidative chain initiation paths, thus reducing the ignition delay time. The end outcome, as corroborated with experimental results, is a faster, more complete and lower temperature combustion process that results in "ultra-clean" exhaust gases.

INTRODUCTION

Of the three basic I.C. engine combustion-ignition types, the spark-ignited gasoline engine and the compression-ignited diesel engine have received the most attention during the previous two centuries. However, the search for a means to simultaneously reduce engine pollution, while also improving both the efficiency and the tolerance of I.C. engines to a wider range of fuels, has recently brought the third basic I.C. engine ignition type to the forefront: the radical-ignited engine. Radical ignition is a form of " homogeneous combustion."

The early work of Noble Lauret Semenov [1] in Russia spawned the first pioneering development in the area of "homogeneous combustion" (or chemical-controlled ignition) by his student, Gussak. This new concept evolved from Gussak's work to further develop an older dual-chamber IC engine type which used the pre-chamber torch-like ignition process [2] first invented by Ricardo [3]. In time Gussak's studies unveiled some unusual phenomena of much greater scientific interest than Ricardo's engine, the pursuit of which ultimately resulted in not only a modified design, but more importantly, a completely new and revolutionary combustion approach [4-5]. The end result was the first radical-ignited engine, the novel low-temperature, high-activity torch ignition engine (LAG engine).

After the LAG engine, the next effort to make use of intermediate species and radicals to induce homogeneous combustion was begun serendipitously at the U.S. Naval Academy while studying the use of acoustic control to increase turbulence in an I.C. engine. In time this pressure wave concept was applied to create a new engine design [6]. The major geometric characteristic of this early design was a second chamber (annulus) on the outside of a straight piston, slightly below its crown. This preliminary design relied on the principle of acoustic resonance and the interaction induced in each of two chambers (the primary combustion and the secondary resonance chamber) to enhance turbulent mixing and combustion. This modified engine could run far leaner with gasoline (air/fuel ratios of 24:1 vs.16:1) than its baseline engine and was found to provide substantially improved performance. Later it was observed that a single cylinder gasoline engine using this piston with annulus design could run without a spark, once started with a spark, if conditions were correct. Preliminary documentation of this finding is given by Allen, et. al. [7] for the transition from spark ignition to gasoline "compression ignition" (CI). It was noted that the "CI" operation was remarkably stable, with very little cycle-to-cycle variation. By using an optical engine and high speed camera, it was also found that a very bright blue emission was detected while approaching "CI" and that, as in the LAG engine, the self ignition "initiated" at many points, instead of a single point (as seen in conventional SI operations). This dramatic change in the combustion characteristics clearly indicated that this piston design involved not only an acoustic phenomenon but also a chemical phenomenon, and that the chemical aspect far outweighed the fluid-mechanic enhancement of the combustion (due to only acoustic enhanced turbulence). This engine design was also later found capable of burning methanol at compression ratios as low as 4:1 without a spark (once started with a spark). Further studies of the acoustic and chemical processes involved in this engine eventually led (through a series of three SCS designs) to the development of the new type of dual-chamber piston geometry design (now called SCS-DI) [8] shown in Figure 1. This most recent member of the SCS family has been found capable of taking diesel stock Euro-II engines to the Euro-III standard by using only piston geometry modifications and injection timing changes.

In general, a homogeneous-combustion engine behaves like a compression-ignited stirred-chemical reactor, with the chemical kinetics of the reacting fuel-air mixture controlled only by temperatures and/or species concentrations. Instead of involving a progression of the combustion process through the charge, as in the older approaches, HC evokes a reaction sequence that involves the simultaneous envelopment of most of (or at least much more of) the charge. Thus, homogeneous combustion results from simultaneous uniform ignition throughout a well-mixed fuel-air charge. Also, HC occurs in relatively lean mixtures and with peak temperatures mostly below the temperature of NO_x formation.

There are several ways available to cause this simultaneous envelopment- all of which better enable the control of the combustion process. The radical-ignition approach makes it possible to have much better control of the combustion process and enables autoignition at low compression ratios. And with significantly greater control of the combustion, the stability and uniformity (and thus also the repeatability from cycle to cycle) of the burn is improved. Also, as will be shown herein, radical induced homogeneous combustion can make feasible the autoignition of low cetane fuels such as methanol at compression ratios that are much lower than normally required for fuels with such poor ignitability. In radical induced HC, NO_x and soot in the exhaust are negligible, providing the incentive for investigators around the world to seek control of the HC process.

In addition to radical ignition (RI), HC has been achieved by compressing a fuel-air charge which has been previously heated electrically in an unthrottled intake manifold *or* heated by exhaust gas either through direct exhaust gas recirculation (EGR) or by exhaust re-admission controlled by variable valve timing, as discussed in [9-11]. Under such strategies, investigators of the HC process have postulated that *full–load* HC is not possible. The authors of reference [9] propose a complex strategy to overcome past problems for direct injected (DI) engines attempting to use HC. However, RI is a much simpler approach to HC and is based on the control of the combustion kinetics.

In the literature TI induced homogeneous combustion in four-stroke gasoline engines is now most commonly referred to as HCCI (homogeneous-charge compression-ignition). For uniformity in nomenclature we now refer to RI induced homogeneous combustion as SCRI (stratified-charge radical-ignition). HCCI is attributed to Najt and Foster [12] who investigated methods of achieving low compression (7.5:1 and 10:1) auto-ignition of 60 and 70 octane gasoline in four stroke cycle engines. Earlier experimental investigations [13] had linked controlled auto-ignition in two stroke cycle gasoline engines to the presence of chemical radical species and to Semenov's work [1] on the low temperature peninsula of radical ignition limits.

The simpler Sonex method of inducing HC through RI avoids thermal heating by mechanically complex preheating and mechanical controls and instead uses available in-cylinder radical chemical species to cause ignition. Thus, it is now understood that homogeneous combustion can be induced by two different methods [14]: RI (radical ignition from the presence of active intermediate species and activated radicals) and TI (thermal ignition from the heat of EGR (in 4 stroke engines) or exhaust carryover (2-stroke engines)). Herein RI (radical ignition) is understood to mean autoignition induced primarily by the presence of these active intermediate species and/or radical species.

The work of [15] suggests that, in some 2-stroke applications involving exhaust gas carry over (from the previous cycle), the higher temperatures of the new (mixed) air charge, coupled with the presence of some unburned fuel (also from the carry-over), causes the generation of radicals during the early part of compression stroke of the next cycle. Thus both thermal and radical effects are involved, with the thermal effects initiating the enhancement process. This would account for the much lower than normal self-ignition temperatures in such engines.

The phenomenological difference between HCCI and SCRI lies in the fact that, in HCCI, the oxygen content of the initial main charge is diluted through the use of inert mass addition (EGR) and its temperature is also increased to cause auto-ignition of a leaner mixture (providing greater homogeneity with respect to temperature and to the oxygen concentration), while, in the second, it is the presence of special homogeneously distributed active intermediate chemical species and radicals in the initial charge that causes auto-ignition at temperatures well below the traditional ("normal") CI temperature for a given fuel. In both cases however, the bulk of the charge ignites all at once (simultaneously or "homogeneously" - thus "homogeneous combustion"), providing faster and more uniform combustion.

It should be noted that combustion in four-stroke diesel engines using EGR is also often labeled HCCI even though the recycled exhaust gases are best cooled prior to their mixing with the inlet air. In this case it is not TI, but the dilution of the charge by inert mass (creating a leanness of oxygen), that enables the compression ignition to be "more" homogeneous in nature. However, in the stricter sense, the combustion in these diesel applications (with EGR) is still via the same stoichiometric diffusion flame process used in regular (i.e., without EGR) diesel combustion applications and these engines do not truly involve homogeneous combustion.

Since the initial work of Najt and Foster [12] on the 4-stroke engine and Noguchi, et. al. [13] on the 2-stroke engine, the rate of work on the use of EGR (in 4-stroke engines) and exhaust gas carry-over (in 2-stroke engines) to achieve homogeneous combustion via HCCI has accelerated [16-30 (four stroke) ; 31-33 (two-stroke)]. Follow-on work to achieve homogeneous combustion by SCRI, has been more limited [34-37].

SONEX SCRI

While the difficulty of controlling the timing of HC ignition was early recognized [7], the invention of a simple in-cylinder control method [39-40] using direct injection and relying on production and retention of chemical intermediate (H_2O_2, CH_2O) and radical (HO_2) species for ignition provided a break-through. When using lower compression ratios, this process for DI engines SONEX Combustion System (SCS) is called SONEX SCRI, or simply SCRI.

The basic idea of SCS engines is to directly control the combustion process inside the combustion chamber itself through modifications to the piston geometry. The SCS-DI engine consists of a dual-chamber system: a primary combustion chamber and a secondary chamber (cavities) built in the piston bowl area, as shown in Figure 1. The primary chamber and the secondary chamber (called the "micro-chamber (MC)") are connected by small passages (or narrow channels). In comparison with the earlier annulus piston design [6], the latest SCS piston geometry (SCS-DI) dampens the acoustic oscillation but preserves the capability of causing the same chemical processes found much earlier [7].

To accomplish radical ignition in Diesel engines the SCS-DI technology integrates several uniquely shaped cavities (micro-chambers, MC's as shown in Figure 1) into the inside of the piston around the circumference of the piston bowl (i.e., embedded into the metal of the piston). These micro-chambers thus form a ring around the piston bowl, with each micro-chamber positioned in line with a spray from the fuel injector. The micro-chambers are connected to the piston bowl by tunnel-like vents (channels or tubes) arranged strategically so that some of the fuel can be trapped in the micro-chambers but also is such a way to insure that the flame from the main chamber is quenched by the vent – preventing complete combustion inside the cavities inside these micro-chambers. Only slow and incomplete oxidation (of the trapped fuel) can take place - resulting in the formation of highly reactive radicals and intermediate species. Thus, in the SCS-DI engine, the primary function of the micro-chamber is to generate the active intermediate and radical species needed for later use in the SCRI combustion process in the main chamber.

The products of the incomplete combustion can then flow out of the cavity (at points in time that depend on the pressure differential), can mix with both the present charge and the next cycle's fresh charge in the main chamber and can thus effect changes in the chemical properties of both charge mixtures to enable RI initially and more complete combustion both initially (during and just after autoignition) and later (in the power stroke). The later effect (in the power stroke of the present cycle) is further enhanced by the intense interaction between the combusting gases of the main chamber charge and the high velocity turbulent jets exiting the micro-chamber [37,41]. These jets carry radicals into the charge of the present cycle and generate intense turbulence. Thus there is a rapid mixing of these radicals and active intermediate species with the charge, and combustion is extended. However, more of the radicals and intermediate reactive species remain trapped in the micro-chamber for use in RI in the next cycle.

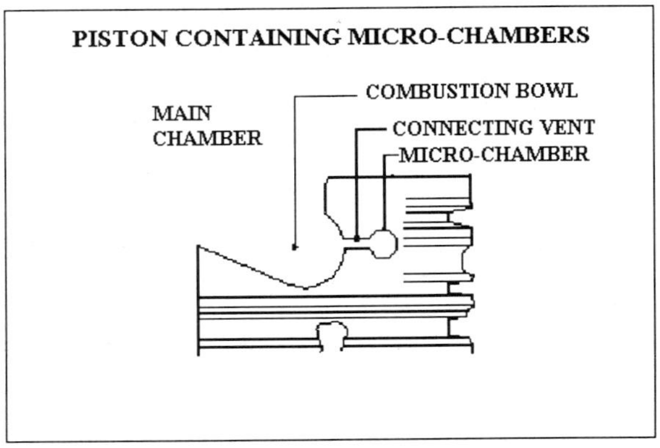

Figure 1 The Sonex Combustion System (SCS) – General Features.

Variants of the SCS-DI design have since been found to enhance combustion in not only 4-stroke ultra-lean SI and DI Diesel engines [40], but also (with some modification) in lightweight 2-stroke engines fitted with spark plugs (to maintain ignition consistency) and operating on JP5 and JP8 fuels [42]. In this two-stroke engine the micro-chamber is shifted from the piston to the inside of the cylinder head. Combustion is sustained for this engine at a 7:1 compression ratio with these higher cetane fuels. This lightweight SI Diesel/JP-5 fuel engine has been deployed since 1998 in ship launched un-manned aerial vehicles (UAV) by the U.S. Navy and the U.S. Marine Corps.

At this point SCS-DI technology branches into two distinctive designs; the one most developed and now in pre-production is called the Low Soot Design for high compression (greater than 16:1) TCI DI diesel engines. The other design called Stratified Charge Radical Ignition (SCRI) for low compression (less than 12.5:1) DI engines is now in transition from a single cylinder NA DI diesel research engine to commercial multi-cylinder turbo-charged engines. For both the Low Soot Design and for SCRI, the MC's produce radical/intermediate chemical species. However, for SCRI there is special design emphasis in this chemical aspect, while for the Low Soot Design (with lower chemical activity), the emphasis is in placing the vent (with its high velocity jet issuing out of it) for maximum interaction with the soot cloud. The Low Soot Design (LSD) is used at normal diesel compression ratios, thus NO_x production is higher than with low compression SCRI. Despite these distinctions, the same phenomena responsible for the unique success of each design are found to a lesser degree in the other design. Thus, so that the process can be fully understood, both designs are described herein.

NUMERICAL STUDIES OF SCRI PHENOMENON AND RESULTS

For SCRI, pistons containing micro-reaction-cavities or chambers as in Figure 1, are designed to produce the required partial oxidization reactions within the MC while retaining sufficient species concentration to seed or fumigate the succeeding engine cycle. With high-speed Schlieren photography in an optical access engine it was observed that all piston crevices - top land, ring, and grooves evacuate the residual within at the beginning of the intake stroke. Thus seeding or fumigation of a new charge can occur by out-gassing from piston micro-chambers.

With SCRI, the start of combustion is controlled by coordination of injection timing to appropriately chosen MC design parameters (volume, location, connecting passage or vent cross section) for a given combustion chamber and engine specifications. Since this process cannot be duplicated in a conventional laboratory constant volume experiment, the verification of the kinetic processes within a MC had to await detailed CHEMKIN kinetics simulation [43] which showed that for stoichiometric H_2-air combustion the intermediate and radical species (H_2O_2, HO_2) produced during compression of are quenched to stable frozen equilibrium during the expansion process in an engine (as shown in Figure 2). If a fraction (typically some 10%) of such species are retained in a MC and re-introduced in the next engine cycle, these species remain stable during compression and become reactive on approaching TDC to initiate ignition.

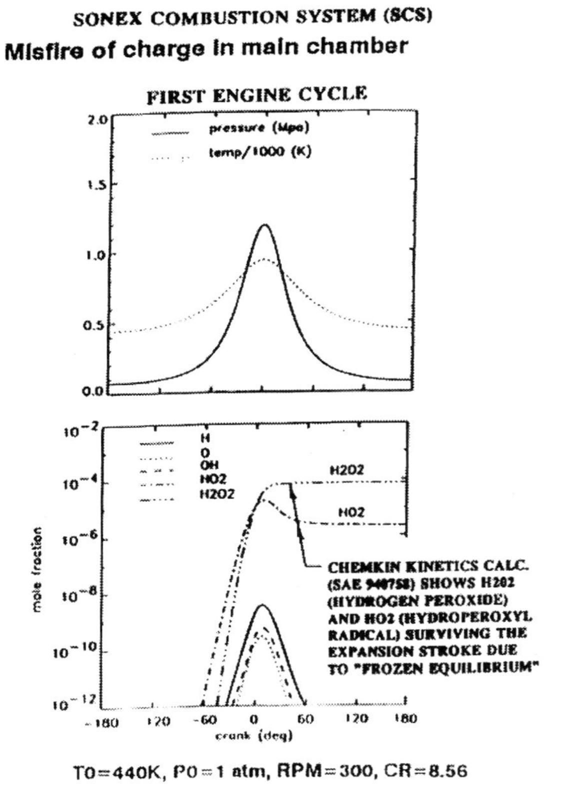

Figure 2 First cycle H_2-air combustion chamber main calculations (upper) and MC calculations (lower).

Shown in Figure 3 are sample calculated results from [43] H_2-air combustion demonstrating this main chamber kinetic process.

A more recent detailed kinetics study of stoichiometric methanol (CH_3OH)-air combustion shows conclusively that H_2O_2 and CH_2O are the dominant intermediate species formed and retained in frozen equilibrium in a MC along with "frozen" HO_2 radicals. This can be clearly seen in the sample micro-chamber calculations for methanol – air combustion given in Figures 4 and 5. After being out-gassed, these species initiate ignition in the combustion bowl by causing major changes in the initiation paths of main chamber oxidative chain reactions. These findings are based on the combined detailed-kinetics calculations of CHEMKIN-II [44] and the multi-dimensional full-simulation calculations of KIVA-II [37,41,45].

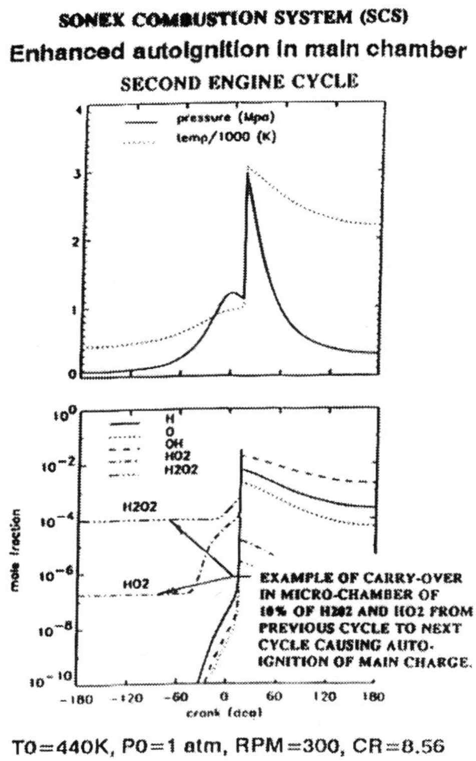

Figure 3 Second cycle H_2-air combustion main chamber calculations (upper) and (lower).

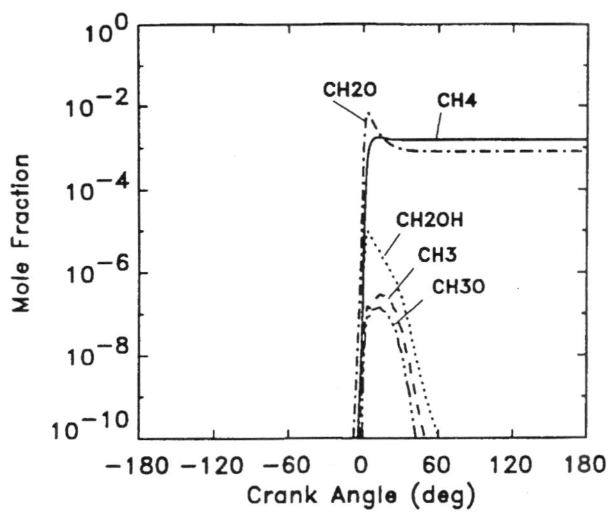

Figure 4. Evolution of CH_2O and Other Selected Species for the Partial Oxidation in the Micro-Chamber of the Stoichiometric CH_3OH-Air Mixture (N=300 RPM, CR = 8.56, $\Phi = 1.0$, SA = 29.0° BTDC, $T_0 = 440$ K, $P_0 = 0.73$ atm).

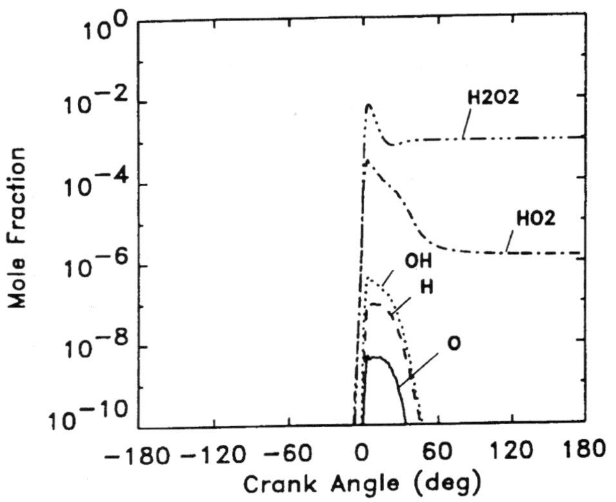

Figure 5. Evolution of H_2O_2, H_2O and Other Selected Species for the Partial Oxidation in the Micro-Chamber of the Stoichiometric CH_3OH-Air Mixture (N=300 RPM, CR = 8.56, Φ = 1.0, SA = 29.0° BTDC, T_0 = 440 K, P_0 = 0.73 atm).

Driven by a set of volume-averaged time-dependent pressures and temperatures obtained from full simulation turbulence combustion calculations performed using KIVA-II, the CHEMKIN-II simulation results given in Figures 4 and 5 were obtained for the stoichiometric methanol-air mixture partial oxidation process in the micro-chamber. As can be clearly seen in these figures, during this process, several stable species, including methane (CH_4), formaldehyde (CH_2O), hydrogen, and hydrogen peroxide (H_2O_2), are formed. Also carbon monoxide (CO), carbon dioxide along with trace amounts of C_2H_6, C_2H_4, C_2H_2, HCO, and HO_2 are found in the unburned mixture (not all shown in Figures 4 and 5). Noticeably, the concentrations of these species become "frozen" (or chemically inert) at different points during the expansion stroke, and remainders of these species are thus available for mixture with the fresh charge of the next cycle. The temperature drop in the micro-chamber, due to the expansion process (during the power stroke) in the main chamber, is what causes the chemical activity of these species to cease ("freeze"), and they remain inert (inactive chemically) at the lower temperatures that prevail until later in the next compression stroke.

A side study was conducted to determine the potential of the most dominant intermediate species (H_2O_2) having a catalytic effect on the ignition of the CH_3OH-air mixture in the main chamber. In this main chamber combustion study (via CHEMKIN-II), the ignition delay for both a baseline engine and an engine with the SCS with an addition to the initial charge of 0.1% H_2O_2 are examined (for a stoichiometric CH_3OH-air mixture at a constant pressure).

As can be seen in Figures 6 and 7, the presence of 0.1% H_2O_2 causes a reduction in the time required for autoignition for this calculation from approximately 38 ms to 15 ms. Similar though less profound, are results obtained using 0.1% CH_2O. As indicated in these results and in an analysis of the rates of reaction for each of the 84 kinetics reactions involved in this simulation: both CH_2O and H_2O_2 can reduce the ignition delay of the CH_3OH-air mixture by changing the dominant chain initiation paths. When 0.1% CH_2O is added instead, the rate of H_2O_2 greatly accelerated. For both, the thermal decomposition of H_2O_2 was found to be the key reaction responsible for rapid hot ignition. A faster build-up of H_2O_2, whether it results from the initial addition of itself or from the initial addition of CH_2O (or the addition of both), results in the significantly shorter ignition induction times.

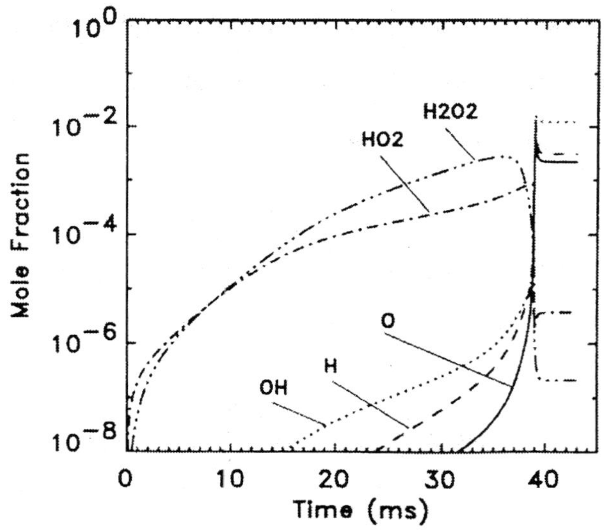

Figure 6. Evolution of H_2O_2, HO_2 and Other Selected Species in the Oxidation in the Main Chamber of the Stoichiometric CH_3OH-Air Mixture; Baseline with P=1 atm, T_0=1000 K.

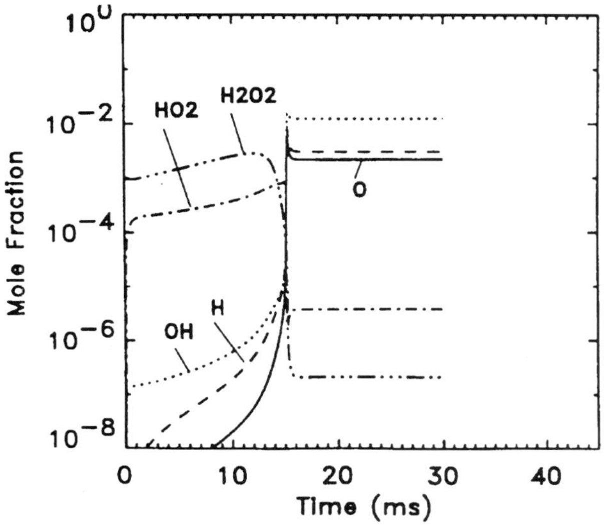

Figure 7. Evolution of H_2O_2, HO_2 and Other Selected Species in the Oxidation in the Main Chamber of the Stoichiometric CH_3OH-Air Mixture with 0.1% H_2O_2; P=1 atm, T_0=1000 .

Under real engine conditions and the assumptions given in [46], a final set of combined KIVA-II and CHEMKIN-II simulations were conducted to study the effect of varying percentages of H_2O_2 carryover to the next cycle. Figure 8 represents the Autoignition Limit of the stoichiometric CH_3OH-air mixture in an engine with a CR of 12:0 and an intake pressure P_0 of one atmosphere. The temperatures on the vertical axis are the minimum intake temperatures found to be required to auto-ignite the charge by compression (radical ignition). For the two RPM values depicted, the regions above the curves are the autoignition regions and the regions below the curves are the misfire regions. In this study, many cases were tried under various conditions. Only for certain cases is the enhanced autoignition predicted. For a compression ratio of 12:0 and RPM of 300, for example, the mixture does not burn spontaneously (i.e., without H_2O_2) until the intake temperature is raised to 433 K or more. With increased addition of H_2O_2 in the mixture, the minimum intake temperature for auto-ignition decreases, due to the catalytic influence of H_2O_2 (as discussed earlier). Also, it is found from these calculations that, for methanol fuel, the concentrations of intermediate species formed in the micro-chamber are strongly dependent on the temperature and engine speed. Moreover, success of autoignition under typical SI engine compression ratios requires sufficient intermediate species and intake temperatures and low enough engine speeds.

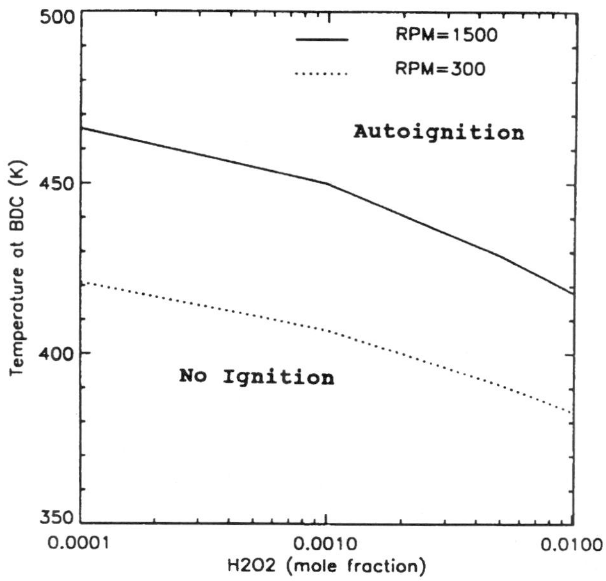

Figure 8. Autoignition Limit of the Stoichiometric CH_3OH-Air Mixture with the Addition of Varying Percentages of H_2O_2 in an Engine with a CR = 12.0 and a P_0 = 1 atm.

These results are indirectly corroborated by the experimental methanol naturally aspirated DI engine studies reported in [34]. With only normal (ambient) air inlet temperatures, for over a 800-2000 range of RPM's, unassisted compression ignition was found to be attainable for full-75% loads in a one-cylinder engine with a 17.5:1 ratio (and at all other load levels with appropriate assist) using the SCS. Otherwise methanol normally requires a much higher compression ratio to autoignite (26:1). Also, using an older version of SCRI [38], a much earlier study [47] was conducted at ambient temperatures on an experimental DI engine. In this study unassisted methanol operation was demonstrated over the entire load range for various RPM's, at a compression ratio of 17:1. Thus, the raising of the compression ratio to diesel levels can lower the required intake temperature for auto-ignition.

These studies illustrate that the main reason for a marked improvement in the burn (and thus pollution reduction) is the homogeneous nature of the auto-ignition process. Only a portion of the gases trapped in the micro-chamber exit during the power stroke. A significant fraction of the active intermediate species and radicals are preserved by design and held over for addition to the new air charge of the next combustion cycle. These species can be retained because their chemical activity becomes "frozen" in the expansion (power) stroke as the pressure (and thus temperature) inside these cavities drop sufficiently low, as shown in Figures 4 and 5. During the exhaust stroke a portion of these species are expelled from the micro-chambers, but many remain. Thus, this remainder mixes with the incoming air charge during the intake stroke. Then, during the compression stroke, as the overall temperature of the air charge increases with crank angle, these species again resume their chemical activity and their presence completely alters the chemical kinetics of the combustion process of the next cycle. Acting somewhat like a catalyst, these radicals and active intermediate species enable autoignition of the main charge at temperatures some 100-200 degrees C below those normally required for compression ignition in Diesel engines. Because of the high mixing levels, this radical initiated autoignition takes place *at many sites throughout the main chamber at the same time.* Thus there is a simultaneous envelopment of most of the fuel/air charge in the cylinder- resulting in a more rapid and more complete burn (compared to normal Diesel combustion) and producing lower peak temperatures. Therefore, not only is there a reduction in soot and unburned hydrocarbons produced in the primary combustion process for diesel fuels, but also there is a significant reduction in nitrogen oxide formation (NO_x) due to the combustion taking place at temperatures mostly below those required for their formation.

RECENT EXPERIMENTAL SCRI STUDIES AND RESULTS

A series of feasibility studies for the U.S. military was completed to examine the ability of the SCS to use JP5/D2 fuels at compression ratios of 9.5:1 and 12.5:1 in a single cylinder (bore 82mm) NA DI (naturally aspirated direct injection) four-stroke cycle engine. In these studies a conventional diesel electrical manifold heater was used for starting. For these studies both the stock mechanical injection system and an electronic fuel injection (EFI) system (still under development) were used. .

Prior to this SCRI feasibility study a DI Diesel engine with MCs in the pistons had previously been designed with priority given to soot reduction (> -60% below baseline, also > -15% NO_x) according the principles in reference [40] and with a compression ratio 18:1. This is now called the SCS Low Soot Design (LSD). This engine, when further converted to SCRI, showed smooth stable operation at all loads and speeds (1000-3500 rpm) at the lower compression ratios (9.5:1,12.5:1) when using the above fuels, accompanied by major improvements in NO_x and soot. For the SCRI design, fuel consumption was better with the stock mechanical system than the EFI (since the EFI is still under development) but rated speed (3500rpm) could be set with EFI due to its greater range in injection timing.

Successful extension of SCRI to turbo-charged engines has also been achieved with fuel economy essentially equal to the original TCI engine and 16:1 compression ratio. For this extension, the rated power of the original was achieved.

Table 1 compares average full to quarter load results for the stock 20:1 engine with stock pistons with the same engine with pistons containing preliminary JP5 design MCs at 2000 rpm using stock mechanical injection system (injection pressure = 400 bar at rated rpm) for the 18:1 *SCS LSD NADI design* and the *12.5:1 SCRI preliminary JP5 design.* The same timing was used in each case, therefore, not the optimum.

Table 1 Emissions (G/kWh) for various single cylinder tests and fuels.

	NO_x	Soot	BSFC	CO	HC	FUEL
20:1 STOCK ENG./With Out EGR	14.9	0.18	377	13.17	6.81*	D2/.05S
18:1 SCS LSD NADI/ /Without EGR	12.5	0.06	387	22.58	5.00*	D2/.05S
12.5:1 SCRI/With Out-EG R	6.98	0.00	565	53.94	4.91	JP5
12.5:1 SCRI/With EGR	2.22	0.01	484	29.75	2.29	JP5
12.5:1 SCRI/With EGR Using EFI	2.75	0.01	384	34.4	0.44	JP5

*Extrapolated

It is important to note in Table 1 that, even with this preliminary piston design for JP5-fuel, there is a *significant reduction* in NO_x and soot for SCRI for the same loads and speeds. Also, it was noted that fuel consumption is greatly affected by: (1) injection timing, which was limited with the stock mechanical injection system, and (2) the manifold temperature, which was 101 deg. F without EGR and 10 to 20 deg. higher with EGR. The increase in CO was noted at all operating conditions but is still comparable with the low CO inherent in DI diesel engines. SCRI is now being transferred to a multi-cylinder TCI engine with the same trends evident as seen in the single cylinder engine but with SCRI fuel economy for the 12.5:1 design equivalent to the normal high compression TCI diesel engine (16:1).

Given in Figures 9-11 are the results for JP5 at a compression ratio of 12.5:1 *with* and *without* EGR, the stock mechanical injection pump and EFI. The JP5 data for smoke, NO_x and fuel consumption given in Table 1 come from the averages for these curves.

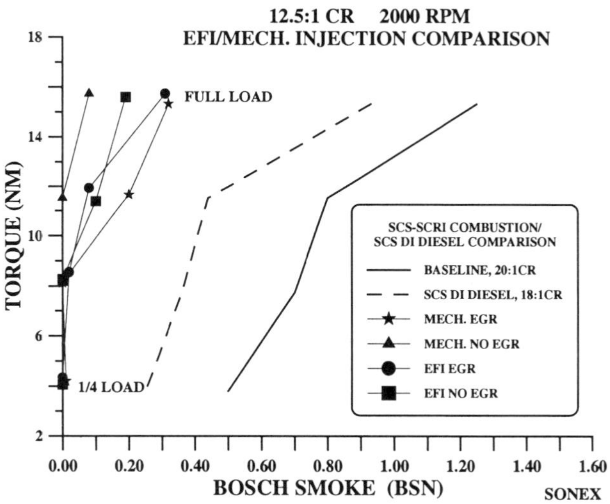

Figure 9 Smoke levels for varying engine loads for various engine configurations.

From the data presented in Table 1 and Figures 9-11 we have identified a significant difference in SCRI from conventional DI diesel combustion at high compression (>16:1): at all loads and speeds at 9.5:1 and 12.5:1 compression ratios, ignition is *single phase*. In all test work over a wide range of loads and speeds, there is no diffusive phase evident in our combustion analyzer results as is shown through the combustion analyzer results given in Figures 12 and 13 for full and ¼ loads, respectively.

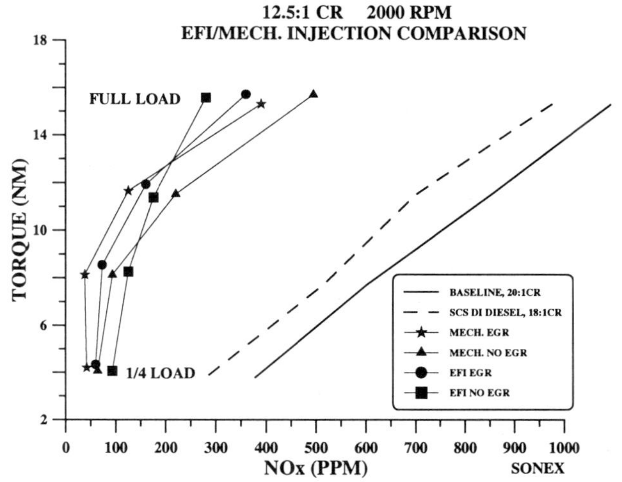

Figure 10 NO_x production for varying engine loads for various engine configurations.

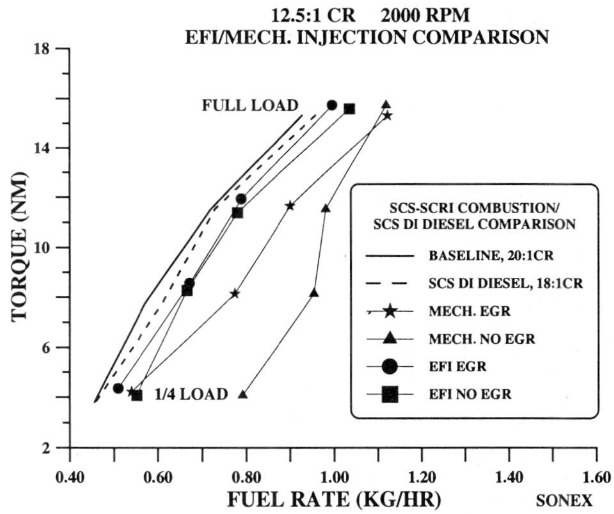

Figure 11 Fuel consumption for varying engine loads for various engine/injection system configurations.

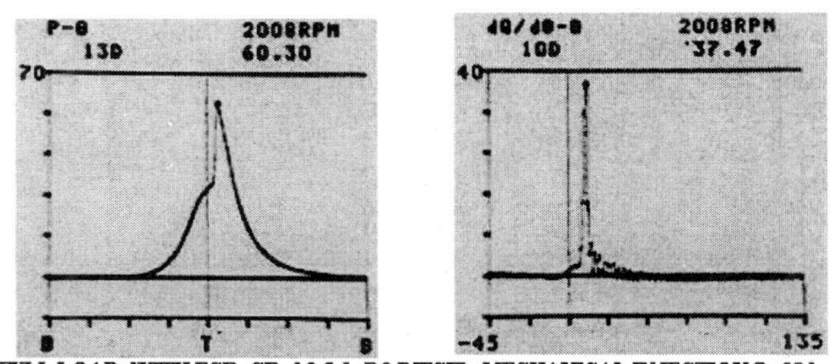

FULL LOAD, WITH EGR, CR=12.5:1, D2 DIESEL, MECHANICAL INJECTION@ -20deg.

Figure 12 Pressure and heat release rates versus crank angle for full load with a 12.5:1 compression ratio SCRI engine burning D2 diesel.

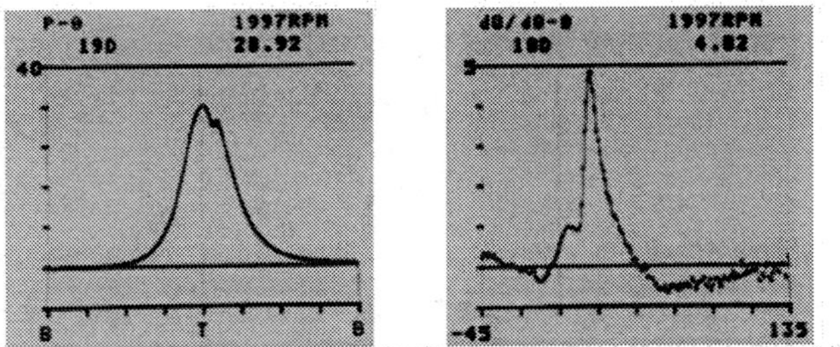

Figure 13 Pressure and heat release rates versus crank angle for 1/4 load with a 12.5:1 compression ratio SCRI engine burning D2 diesel.

DISCUSSION OF RESULTS AND CONCLUSIONS

From the combination of the results presented herein one can clearly observe several of the basic chemical-kinetics mechanisms responsible for enhanced auto-ignition in the SCS. For methanol fuel, the species CH_2O, H_2O_2 and HO_2 (formed during the low-temperature partial oxidation process in the micro-chamber) remain stable ("in frozen equilibrium") during the later part of the expansion stroke and all of the exhaust and intake strokes and slowly mix with the main charge. The first two of these species are expected to be primarily responsible for the significant results presented for the larger hydrocarbon molecule fuel SCS experiments presented (for JP5 fuel). These species are shown to be retained in sufficient quantities in methanol combustion to be able to change the dominant oxidative chain initiation paths of the auto-ignition process, substantially reducing ignition delay and enabling hot-ignition to begin at much lower temperatures. The homogeneous-combustion nature of the process enables the burn to be more rapid and more complete, reducing CO and unburned hydrocarbon emissions. The lower-temperature combustion process in the main chamber, in turn, reduces NO_x production. This follows from the direct relationship between NO_x production and peak burned gas temperatures [14].

Though H_2O_2 and CH_2O were found present in larger quantities in the SCS, the HO_2 radical, though present in much lower quantities is also seen to act as a chain prompt, and higher concentrations of HO_2 accelerate this process. Thus, in the low soot version of the SCS, during the period of time when the high speed turbulent jet is transporting micro-chamber partial oxidation products into the main chamber (after approximately 27° ATDC [35]), the HO_2 build-up in the micro-chamber is significant (as shown in Figure 5 herein) - accounting in part for the much larger reductions in emissions in the SCS when the design is altered to emphasize this part of the technology. Thus in methanol combustion, though it may be that the two dominant intermediate species H_2O_2 and CH_2O are more important because they are generated in higher quantities in the micro-chamber, radical enhancement (via HO_2) of both the main chamber autoignition and the ongoing main chamber combustion process is also certainly a significant a factor. The mechanism by which HO_2 helps augment combustion (via the high speed turbulent jet transport and mixing process) needs additional study however.

Over twelve years of continual experimental work and theoretical study has been required before the chemically enhanced low-temperature auto-ignition phenomena of the current SCS could be adequately understood. Due to this experimental work [7,34,38-40,etc.], the computational work of [14,37,41,43,46] and the present work, the overall operating mechanism of the SCS–DI engine can now be summarized as follows:

Combustion is started via a spark plug or manifold flame glow plug for the first cycle of the engine.

The flame generated in the primary (main) chamber is quenched (due to heat loss) inside the narrow passages/channels leading to the micro-chamber.

During the earlier part of the power (expansion) stroke, unburned fuel-air mixture trapped in the secondary (micro) chamber undergoes a series of partial oxidation reactions at temperatures well below auto-ignition limits. Intermediate species and radicals are formed during this non-explosive luminous exothermic process (usually a cool flame).

Later in the power stroke, intense turbulence, generated by the mass exchange via high-speed jets between the micro-chamber and the main chamber, enhances the rate and completeness of the main chamber combustion process. These jets originate from pressure build-up due to the partial oxidation process in the micro-chamber and expansion in the main chamber. In diesel fuel applications, the oxygen, activated species and radicals carried by these jets collide and rapidly mix with the soot clouds formed in the main chamber- causing a further oxidation of the soot. In gasoline engine applications these same species laden turbulent jets help insure oxidation of CO and unburned hydrocarbons still resident in the main charge.

The restricted inflow and outflow processes of the secondary chamber (cavities) during the later expansion and exhaust strokes allow adequate retention of the partially oxidized micro-chamber charge for later seeding, doping or fumigating of the next intake charge with CH_2O, H_2O_2, HO_2, etc., during the subsequent intake stroke (as observed in optical studies). These species become inactive ("frozen" due to the temperature/pressure drop) during the later part of the expansion stroke and remain inactive until their temperatures are adequately raised during the compression stroke.

The catalytic effect of these homogeneously dispersed intermediate species and radicals lead to enhanced autoignition of the next charge in the main chamber at lower pressures and temperatures than predicted theoretically in the absence of such species (100-200 K below the temperatures required for normal CI). The autoignition takes place homogeneously: i.e., the bulk charge ignites simultaneously and uniformly. Also, as in the LAG engine, this low temperature auto-ignition process is based on the Semenov's Peninsula of ignition limit [1].

These observations are very helpful in enhancing our understanding of the mechanism of the radical and active intermediate species induced low-temperature homogeneous-combustion process, although many aspects of this unfamiliar combustion phenomenon still require clarification. Much more importantly, these findings point to a new concept (or engine design variable) that can be converted into feasible technology for the better control and enhancement of combustion in I.C. engines using a wide variety of fuels.

REFERENCES

1. Semenov, N.N., Some Problems in Chemical Kinetics and Reactivity, Princeton University Press, 1958.

2. Date, T., Yagi, S., Ishizuya, A.and Fujji, I., "Research and Development of the Honda CVCC Engine", SAE Paper 740605, 1974.

3. Ricardo, H.C., "Recent Work on the Internal Combustion Engine", SAE Paper 1922.

4. Gussak, L.A., " High Chemical Activity of Incomplete Combustion Products and a Method of Prechamber Torch Ignition for Avalanche Activation of Combustion in Internal Combustion Engines", SAE Paper 750890, 1975.

5. Gussak, L.A. and Turkish, M.C., "LAG-Process of Combustion and Its Application in Automobile Gasoline Engines", Proc. IMechE,C257/76, 1976.

6. Pouring, A.A., Failla, C.C., Rankin, B.H., Keating E..L., and Riddell, F., "Parametric Variation of a Heat Balanced Engine", Fluid Mechanics of Combustion Systems (Morel, T., Lohman, R.P. and Rackey, J.M., eds.), New York:ASME, pp.41-50, June,1981.

7. Allen, J., Pouring, A. A., Keating, E.L., "Heat Balanced IC Engine Transition Studies", AIAA Paper 82-1116, 1982.

8. Pouring, A.A., Leto di Priolo, C., Combustion Chamber for Direct Injected Reciprocating Piston Internal Combustion Engine, US. Patent 5,862,788, Jan. 26, 1999.

9. Willand, J., Nieberding, R-G, Vent, G., Enderle, C., The Knocking Syndrome: its cure and Potential. Auto. Engr. International, SAE Dec 1999.

10. Ryan, T. W. III, and Callahan, T.J., Homogeneous Charge Compression Ignition of Diesel Fuel, SAE Paper 961160, 1996.

11. Odaka, M, Suzuki, H., Koike, J., and Ishii, H. Search for Optimizing Control Method of Homogeneous Charge Diesel Combustion, AE 1999-01-0184.

12. Najt, P.M. and Foster, D.E., Compression - Ignited Homogeneous Charge Combustion, SAE 830264.

13. Noguchi, M., et al. A Study on Gasoline Engine Combustion by Observation of Intermediate Reactive Products during Combustion, SAE 790840.

14. Blank, D.A., Pouring, A.,A and Lu, J.. "Kinetic Mechanism and Potential of Exhaust Gas Radical Enhanced Autoignition and NO_x Reduction in Methanol and Hydrogen I.C. Engines" (accepted as an SAE Paper).

15. Maji, S., Abraham, M., Pundir, B.P. and Radzimirski, S., "Abnormal Combustion in a Two-Stroke Spark Ignited Methanol Engine", SAE Paper 880174, 1988.

16. Gray, A.B. and Ryan, T.W., "Homogeneous Charge Compression Ignition (HCCI) of Diesel Fuel", SAE Paper 971676, 1997.

17. Hashizume, T., Miyamoto, T., Akagawa, H., Tsujimura, K., "Combustion and Emission Characteristics of Multiple Stage Diesel Combustion", SAE Paper 980505, 1998.

18. Suzuki, H., Koike, N. and Odaka, M. "Combustion Control Method of Homogeneous Charge Diesel Engines", SAE Paper 980531, 1998.

19. Zelenka, P., Aufinger, H., Reczek, W. and Cartellieri, W., "Cooled EGR – A Key Technology for Future Efficient HD Diesels", SAE Paper 980190, 1998.

20. Christensen, M. and Johansson, B., "Homogeneous Charge Compression Ignition with Water Injection", SAE Paper 1999-01-0182, 1999.

21. Seko, T. and Kuroda, E., "Methanol Lean Burn in an Auto-Ignition Engine", SAE Paper 980531, 1998.

22. Harada, A. Shimazaki, N., Sasaki, S., Miyamoto, T., Akagawa, H. and Tsujimura, K., "The Effects of Mixture Formation on Premixed Lean Diesel Combustion", SAE Paper 980533, 1998.

23. Dickey, D.W., Ryan, T.W. and Matheaus, A.C. "NO_x Control in Heavy-Duty Diesel Engines – What is the Limit?", SAE Paper 980174, 1998.

24. Ladommatos, N., Abdelhalim, S.M., Zhao, H. and Hu, Z., "Effects of EGR on Heat Release in Diesel Combustion", SAE Paper 980184, 1998.

25. Ogawa, H., Chenyu, L., Tosaka, S., Fujiwara, Y. and Miyamoto, N., "Combustion Analyses with Inter-Chamber Gas Composition Measurements in a Premixed Lean Compression Ignition Engine", Comodia 98, Kyoto, Japan, 1998.

26. Furutani, M, Ohta, Y., Kono, M, Hasegawa,M, "An Ultra Lean Premixed Compression-Ignition Engine Concept and its Characteristics", Comodia 98, Kyoto, Japan, 1998.

27. Takatsuto, R., Igarashi, T. and Lida, N., "Auto Ignition and Combustion of DME and n-Butane/Air Mixtures in Homogeneous Charge Compression Ignition Engine, Comodia 98, Kyoto, Japan, 1998.

28. Odaka, M., Suzuki, H., Koike, N. and Ishii, H., "Search for Optimizing Control Method of Homogeneous Charge Diesel Combustion", SAE Paper 1999-01-0184, 1999.

29. Flowers,D.,Aceves, S., Smith, R.,Torres, J., Girard, J. and Dibble, R., "HCCI in a CFR Engine: Experiments and Detailed Kinetic Modeling", SAE Paper 2000-01-0328, 2000.

30. Fiveland, S.B. and Assanis, D.N., "A Four-Stroke Homogeneous Charge Compression Ignition Engine Simulation for Combustion and Performance Studies", SAE Paper 2000-01-0332, 2000.

31. Ashley, S., "A Radical way to Burn", Mechanical Engineering pp. 64-67, August 1996.

32. Ishibashi, Y. and Asai, M., "A Low Pressure Pneumatic Direct Injection Two-Stroke Engine by Activated Radical Combustion Concept", SAE Paper 980757, 1998.

33. Duret, P. and Venturi, S., "Automotive Calibration of the IAPAC Fluid Dynamically Controlled Two-Stroke Combustion Process", SAE Paper 960363, 1996.

34. Lu, J. and Pouring, A.A., "Development of a New Concept Piston for Alcohol Fuel Use in a CI Engine", SAE Paper 961078, 1996.

35. Murase, E. and Hanada, K., "Enhancement of Combustion by Injection of Radicals", SAE Paper 2000-01-0194, 2000.

36. Hayashi, K., Matsuura, K. and Baba, S., "Performance of a Flame Jet Ignition System in a Two-Stroke Engine", SAE Paper 2000-01-0199, 2000.

37. Blank, D.A., Pouring, A.A. and Lu, J., "Qualitative Flow Field Studies of Combustion in I.C. Engines Using a Simplified Sonex Bowl-in-Piston Geometry", SIAT/SAE 2001, Pune,

India, 10-13 Jan. 2001, 11 pages (SAE Paper 2001-01-0021).

38. Failla, C.C. et al., Piston and Process For Achieving Controlled Ignition and Combustion of Hydrocarbon Fuels in IC Engines by Generation and Management of Fuel Radical Species, U.S. Patent 4,898,135, 6 Feb. 1990.

39. di Priolo et al., Combustion Chamber for Internal Combustion Engine and Process of Combustion Using Fuel Radical Species, U.S. Patent 5,322,042, 21 June 1994.

40. Pouring, A.A., Leto di Priolo, C., Combustion Chamber for Direct Injected Reciprocating Piston Internal Combustion Engine, US. Patent 5,862,788, Jan. 26, 1999.

41. Blank, D.A., Pouring, A.A. and Lu, J., "Flame Quenching in the Micro-Chamber Passage of I.C. Engines With Regular-Symmetric Sonex Piston Geometry" (accepted as an SAE Paper).

42. McCowan, W.P., Bopp, B.R., Pouring, A.A., Mills, P.Y., Charge Conditioning Systems for Enabling Cold Starting and Running of Spark-Ignited Diesel Fueled Piston Engines. U.S. Patent 5,855,192, Jan. 5, 1999.

43. Lu, J. and Pouring, A.A., A Preliminary Study of Chemically Enhanced Autoignition in an Internal Combustion Engine, SAE Paper 940758, 1994.

44. Kee, R.J., Rupley, F.M. and Miller, J.A.:"CHEMKIN-II: A Fortran Chemical Kinetics Package for the Analysis of Gas-Phase Chemical Kinetics", Sandia National Lab Report, SAND89-80098, UC-706, Nov., 1991.

45. Amsden, A.A., O'Rourke, P.J. and Butler, T.D., "KIVA-II: A Computer Program for Chemically reactive Flows with Sprays", Los Alamos Lab, LA-11560-MS, May 1989.

46. Blank, D.A., Pouring, A.A. and Lu, J., "Methanol Combustion in Low Compression Ratio D.I. Engines Enabled by Sonex Piston Design", SAE Paper 2001-01-1197,2001.

47. Pouring, A.A., "Chemical Acoustic Charge Conditioning for Low Emission I.C. Engines", Vol. I, First International Conference on Combustion Technologies for a Clean Environment, Vilamoura, Portugal, 3-6 Sept. 1991.

NOMENCLATURE

Symbols

N	Engine Speed
P	Pressure
T	Temperature

Abbreviations

A/F	Air/Fuel ratio
ATDC	After TDC
BDC	Bottom-Dead-Center
BTDC	Before TDC
CA	Crank Angle
CR	Compression Ratio
EGR	Exhaust Gas Recirculation
RPM	Revolutions Per Minute
SA	Spark Advance
TDC	Top-Dead-Center

Greek Symbols

Φ	Fuel/Air equivalence ratio

Subscripts

0	Initial state

A New Generation of Engine Combustion Processes for the Future?
© P. Duret (Editor) and Editions Technip, Paris, 2001, pp. 141 - 152
27 rue Ginoux, 75015 Paris

ACTIVATED RADICAL COMBUSTION IN A HIGH-SPEED HIGH-POWER PNEUMATIC DIRECT INJECTION TWO-STROKE ENGINE

Yoïchi Ishibashi, Kenji Nishida, Masahiro Asaï
Honda R&D Co., Ltd.

ABSTRACT

A higher specific power output has been one of the most primitive demands for the vehicle engines. It would not be an exception even for the environmentally conscious engines. Our previous papers proposed a clean engine concept standing on the Activated Radical (AR) combustion and a pneumatic direct injection (PDI). One of the peculiarities of the concept is seen in the light load range, where the controlled auto-ignition copes with the irregular combustion, instead of having recourse to the stratified charge. Which should very much facilitate the in-cylinder mixture formation. A proto-engine based on this concept has ever demonstrated its promising environmentally conscious potential, however the specific power output was not very sufficient for the recent commercial demands.

This time, intending to comply with the higher output demand, another proto-engine has been newly made. The principal modifications of the engine were the addition of a reciprocating pump for the PDI and the specifications for a typical high-speed engine. This engine successfully generated a double time specific power output comparing to the former one. However, at the same time, our chief concern was whether the AR combustion would be applicable to this kind of PDI engine. In order to confirm our PDI-AR combustion concept in the high-speed high-power engine, this paper discusses the effects of these modifications on the AR combustion.

The experimental result showed that the AR combustion speed, which has been a question, increased according to the engine speed. This made it possible for the combustion to take place in a specific crank timing regardlessly of the engine speed. Meanwhile, the attachment of the air pump affected the AR combustion both negatively and positively. The assist air quantity required for the PDI was roughly proportional to the power output, therefore the air pump had to be throttled at the part load in order to maintain the original performances of the AR combustion. Finally it was revealed that our PDI-AR combustion concept was applicable in this pneumatic direct-injection engine which generated a specific power output of 180 kW/ℓ at 11,000 r/min.

INTRODUCTION

Direct injection technology successfully solved the typical drawbacks of the conventional two-stroke engines and several new generation two-stroke engines have been developed [1-4] with demonstrating their potential as an environmentally conscious clean power unit. Meanwhile in a stage of the practical use of these engines, demands on them are gradually escalating. Especially in the case of engines for the fun vehicles such as snowmobiles, personal watercrafts or motorcycles, a higher power output is one of the most primitive demands.

In the direct injection engines, the power output potential depends on not only the engine's charging efficiency characteristic but the fuèl injector potential itself, i.e. the power output is restricted by the maximum injection quantity for a quality mixture performed in a unit time. By increasing the area of the injector nozzle, the maximum injection quantity would theoretically increase, there is however another restriction from the minimum injection quantity. Consequently each injector has its typical dynamic range. The dynamic range would cover approximately 700 kPa of the mean effective pressure's range at 9,000 r/min in the most sophisticated stratified charge engines' case. This becomes 100 kW/ℓ, however, yet it would not be very sufficient for the fun vehicles.

Our previous papers [5, 6] have proposed the PDI-AR combustion engine, whose concept had no necessity for the stratification in the mixture formation at the light load range. From this viewpoint, the dynamic range of the injector should be potentially wider, however its specific power output had not exceeded 80 kW/ℓ. Intending to confirm our PDI-AR combustion engine concept in a higher power output engine, an experimental PDI engine was newly developed. This engine with a 250 cm^3 single cylinder generated a specific power output of 180 kW/ℓ at 11,000 r/min. From the power output demands, the experimental engine has two principal features. One is a reciprocating air pump. It was revealed that the required amount of air for a PDI event was roughly proportional to the engine's power output. In the case of our PDI system, it has reached a 10% volume of the displacement at the maximum power point. It was hence difficult to ensure a sufficient air supply from the compressed in-cylinder gas at the very high engine speed region, independently from the gas exchanging state. The other feature is a special design around the exhaust port and the exhaust valve. The exhaust port, which consists of three passages, provides a wider opening duration and a larger opening area. It was a typical high-speed engine's exhaust port, originated from a world GP racer's engine. These features were however indispensable for the PDI engine obtaining a higher power output, i.e. a higher torque at a higher engine speed, the influence of them upon the AR combustion was our chief concern. Whether the AR combustion would be applicable to this kind of engine, several basic questions came up.

In the case of the SI engine, the combustion speed increases as the engine speed increases. Consequently, the combustion takes place at a specific crank timing regardlessly of the engine speed. This is one of the most beautiful features of the SI engines, which makes them applicable to the high-speed operation. It is explained that the flame propagation speeds up due to the increase of the turbulence in the combustion chamber. Would this desirable feature be transferred to the auto-ignition combustion that have a quite different process? Our previous paper [7] simply explained the onset of the AR combustion by using in-cylinder gas temperature at the end of the compression stroke as a principal factor. The in-cylinder gas temperature was given from the enthalpy balance during the gas exchanging process. According to this theory, the enthalpy of the additional air supplied by the pump should not be negligible for the total enthalpy balance. This paper will discuss the applicability of the AR combustion to the high-speed high-power PDI engine while examining basic questions such as:

1) Would the auto-ignition be controllable even in a very high-speed operation? On what does the combustion speed depend?

2) Does a quantity of the PDI air affect the AR combustion zone to reduce or enlarge?

3) Does the additional friction affect of the air pump reduce the lower limit of the AR combustion?

4) In this case, would the merits of the AR combustion be sufficiently maintained?

EXPERIMENTAL ENGINE AND PRI-TESTS

EXPERIMENTAL ENGINE

The experimental engine has been basically designed according to the concept of the PDI-AR combustion engine. Hence it is equipped with a PDI system and an exhaust valve for the AR combustion control. The mechanism itself has not very distinctive difference from the first prototype. Figure 1 shows the outline of the engine. There are no more channels connecting the air chamber and the cylinder via the rotary valve, because the compressed air is supplied directly into the air chamber by using a pump. The ports for the gas-exchange such as the intake, the scavenging and the exhaust have been designed to give sufficient area to generate over 47 kW at 11,000 r/min. An expansion-chamber was applied for the exhaust system. Principal specifications of this experimental engine are listed in Table 1, being compared with those of the first prototype.

Figure 2 explains the structure of the exhaust valve. The valve made by a pressed stainless plate covers the three exhaust ports at the periphery and the both sides of the valve.

This valve varies the effective opening area of the exhaust port from a 100 % to 5 % at the fully closed valve position. Again the periphery part of the valve plays a role of the exhaust timing control valve as well.

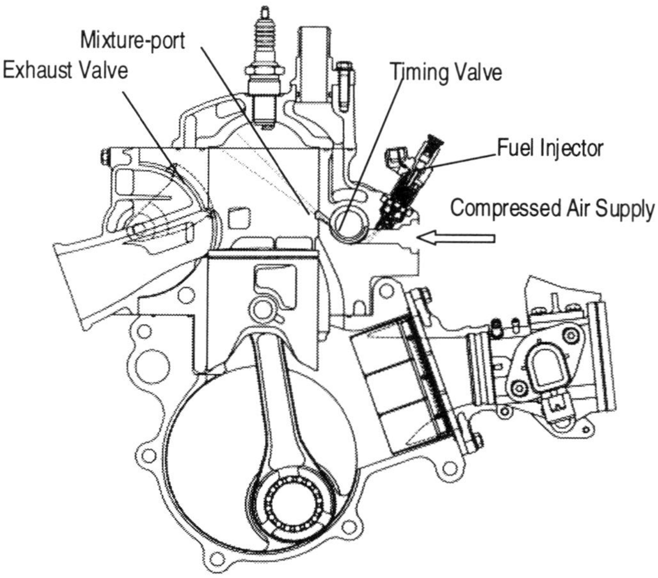

Figure 1. Overview of the experimental engine.

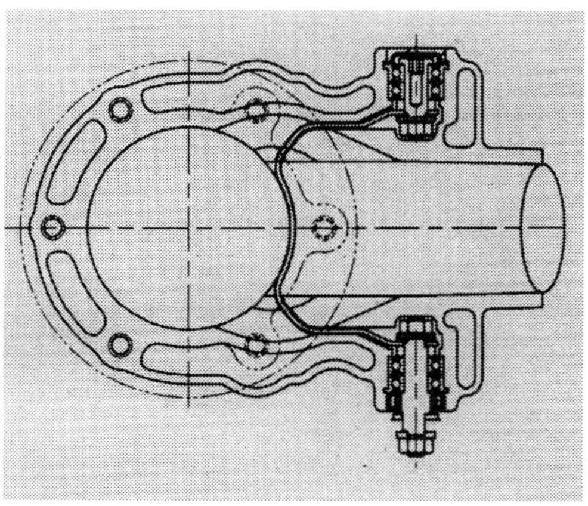

Figure 2. The exhaust valve structure.

Table 1. Principal specifications of the experimental engine

Specifications	Experimental Engine	First Proto-Engine
Bore x Stroke	68 mm x 68.8 mm	80 mm x 80 mm
Displacement	249.9 cm^3	402 cm^3
Compression Ratio (Trapped)	7.2	6.3
Compression Ratio (Geometric)	12.4	10.5
Exhaust Opening Duration	194 deg. C.A.	188 deg. C.A.
Exhaust Opening Area	2437 cm^2 deg.	2159 cm^2 deg.
Scavenging Opening Duration	130 deg. C.A.	113 deg. C.A.
Scavenging Opening Area	1712 cm^2 deg.	1449 cm^2 deg.
Intake Bore	38 mm	38 mm
Exhaust Opening Rate	100% – 5%	100% - 5%

AIR-PUMP SELECTION AND PRI-TESTS

The PDI system was firstly operated by using the laboratory supplied air in order to estimate the required pump capacity. Quantity of the air supplied for the PDI is expressed by using a ratio in this paper. A symbol "Las" is a ratio of the mass of air supplied in a cycle against the reference mass required to fill the displacement under the reference conditions, as the manner of the delivery ratio. The experimental result revealed that the Las required for the PDI was roughly proportional to the specific power output as shown in figure 3.

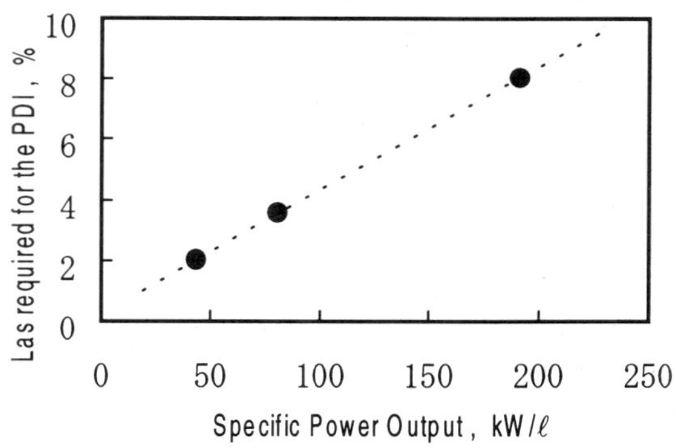

Figure 3. Las required for the PDI.

Thus, for our intention, the air pump should supply every cycle approximately 20 cm^3 of air in terms of the atmospheric conditions. A 63 cm^3 reciprocating air pump shown in figure 4 was chosen considering some margins. The crankshaft drives this pump in the speed ratio of 1 to 1. The liquid cooled cylinder and the oil bath system were applied in order to ensure the tribology at the very high-speed operation. Figure 5 shows the pumping characteristics in combination with the experimental engine. The frictional loss of the pump is represented in the terms of the 250 cm^3 engine's frictional mean effective pressure (FMEP). As shown in figure 5, there was a tendency to increased the Las as the engine speed decreased. This was very natural for the volumetric type air pump, however as figure 3 showing, the required Las was contrary to this characteristic. Therefore, a throttle valve was installed in the inlet of the pump in order to control the pumping work. With this air pump, the PDI system of the experimental engine satisfactorily functioned throughout the revolution range up to 12,000 r/min. The maximum power finally reached 45 kW at 11,000 r/min while driving the air pump.

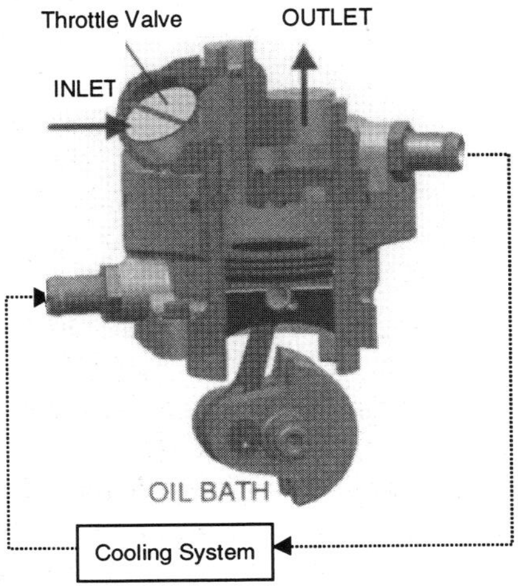

Figure 4. The air pump.

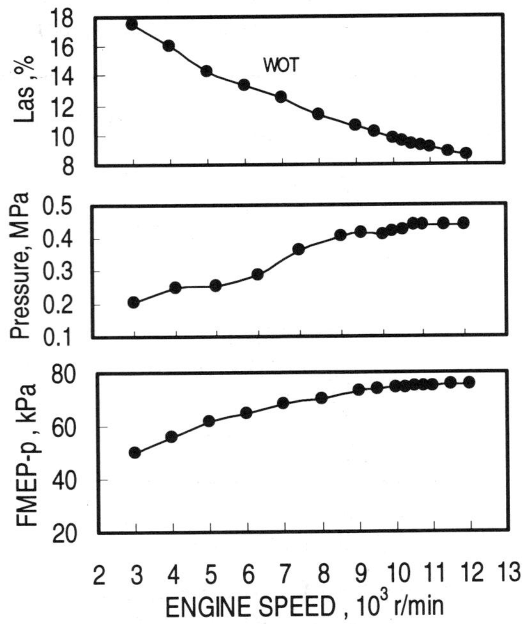

Figure 5. The pumping characteristics of the air pump.

Ultimately, the air supplied for the PDI took part in the fresh charge, hence the additional friction owing to the air pump affected the total performance a little at the wide open throttle (WOT) condition. Again, by giving a proper combination of the exhaust valve opening and the pump-throttle opening, the AR combustion was observed at a brake mean effective pressure (BMEP) – engine speed zone shown in the figure 6 in comparison with our first proto-engine.

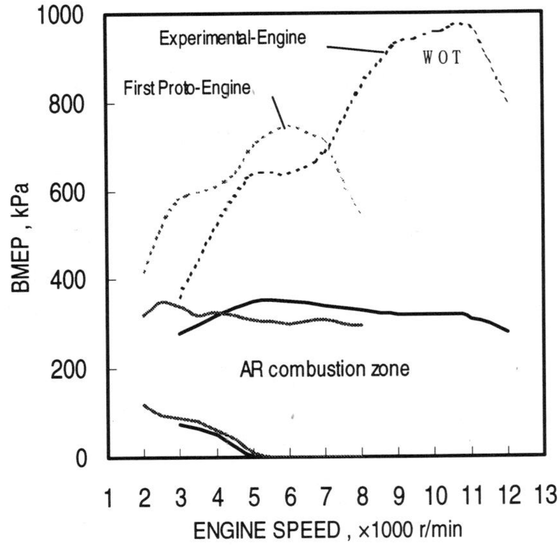

Figure 6. The AR combustion zone.

ACTIVATED RADICAL COMBUSTION ANALYSIS

AUTO-IGNITION COMBUSTION SPEED

Timing controlling mechanism of the AR combustion has been already explained from the gas exchanging aspect [8] and the gas temperature aspect [7], however concerning the combustion speed, it has not been satisfactorily analyzed yet. In the case of SI combustion, the combustion speed depends on the flame propagation speed. Then on what would the auto-ignition speed depend?

Effect of the load

Figure 7 compares the combustion speed of the experimental engine in various loads at 3,000 r/min. As decreasing load, i.e. decreasing indicated mean effective pressure (IMEP), the combustion speed represented as an inclination of the heat release curve generally decreases. It could be explained as the increase of inert gases due to a deterioration of the scavenging efficiency, or as the decrease of in-cylinder turbulence. The combustion type also varies among these curves of the heat release in this graph. The upper two curves were the SI combustion. The third curve from the top was the AR combustion with an assist of the spark ignition, namely semi-AR combustion. The lower four curves were the AR combustion. Focusing the switching timing between SI and AR combustion, this graph shows that the combustion speed once increases and again decreases as the load decreases.

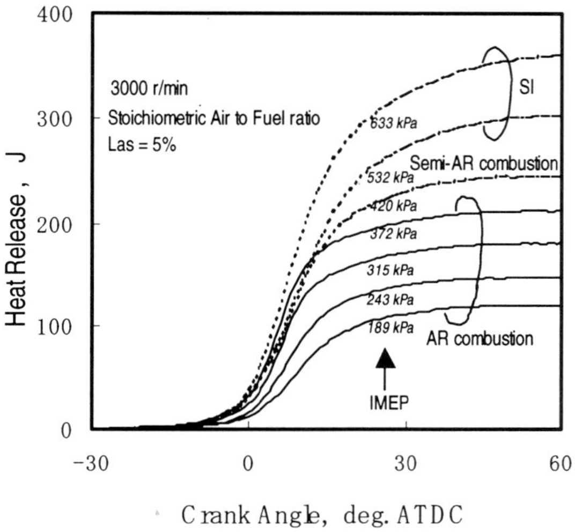

Figure 7. Combustion speed in various loads.

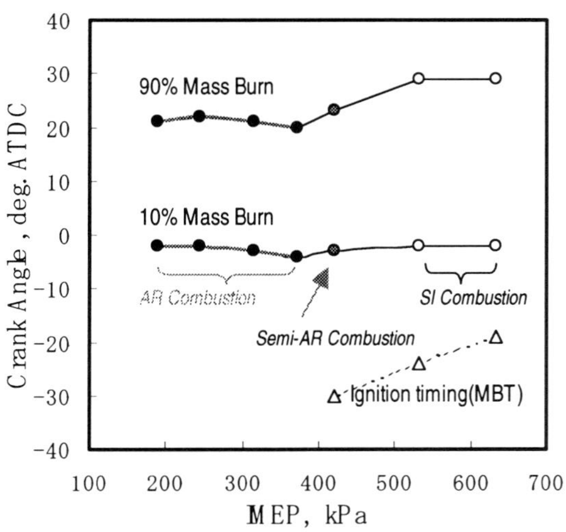

Figure 8. Combustion duration on the crank angle against IMEP.

Again, observing this phenomenon from the mass burn fraction viewpoint, figure 8 shows that the combustion duration rated from 10% to 90% mass burn took place at a constant timing of the crank angle regardlessly of the load in each type of combustion. It could explain that when the combustion speed decreases, the total mass to burn also decreases, thus the combustion duration on the crank angle remains constant.

Effect of the compression ratio

Our previous paper [8] reported that the compression ratio was not a strong factor for the onset of the AR combustion. It was reasoned that a deterioration of the exhaust gas temperature due to the larger expansion ratio canceled the compression heating effect. It has been an experimental result from a comparison between the ordinal compression ratios. However it turned out that an impracticably higher compression ratio enhanced the onset of the auto-ignition. Figure 9 compares the AR combustion at the same IMEP in different compression ratios. The combustion speed is obviously faster in the case of the geometric compression ratio (CRg) of 15 than CRg of 10.9 case. In these cases, there was a grave difference in the exhaust valve opening for the AR combustion. The exhaust valve opening required for the CRg of 15 was 38%, while it was 82% for the CRg of 10.9. In this comparison, the chief factor for the difference in the combustion speed would be a concentration of the inert gases in the cylinder.

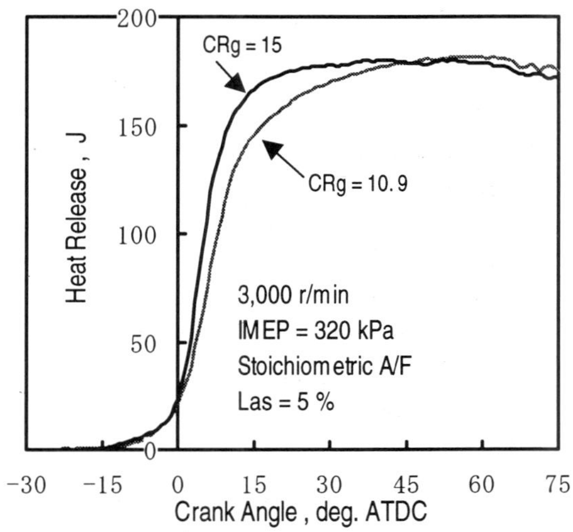

Figure 9. Effect of the compression ratio on the AR combustion speed.

Effect of the engine speed

If the onset timing of the AR combustion would be given by the in-cylinder temperature resulted from the gas exchange process, the scavenging efficiency should be nearly constant in the same IMEP condition regardlessly of the engine speed. The former test results suggest that the concentration of the in-cylinder inert gases is the primary factor for the AR combustion speed. If it be so, the AR combustion speed would be almost constant throughout the rpm range and the combustion might take place until later on the crank angle at a high speed operation of the engine. This is inconvenient for the high-speed engine application of the AR combustion. Figure 10 shows the heat release speed of the AR combustion at a constant IMEP at various engine speeds. Showing the same heat release just before the TDC, the curves become steeper as the engine speed increases. The horizontal axis of the graph is the time from the TDC, thus an incline of the curve represents the combustion speed. Hence the data shows that the AR combustion speed increases as the engine speed increases. Figure 11 indicates that the combustion duration rated by 10% and 90% mass burn timing comes in the nearly same crank angle duration regardlessly of the engine speed. It is the same manner as the SI combustion.

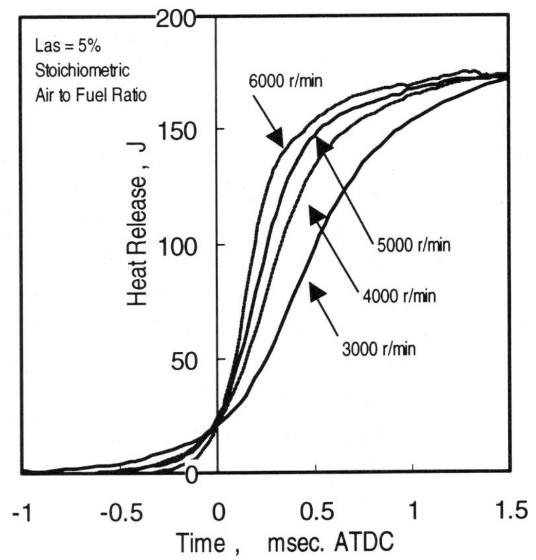

Figure 10. Heat release speed of the AR combustion in the various engine speeds.

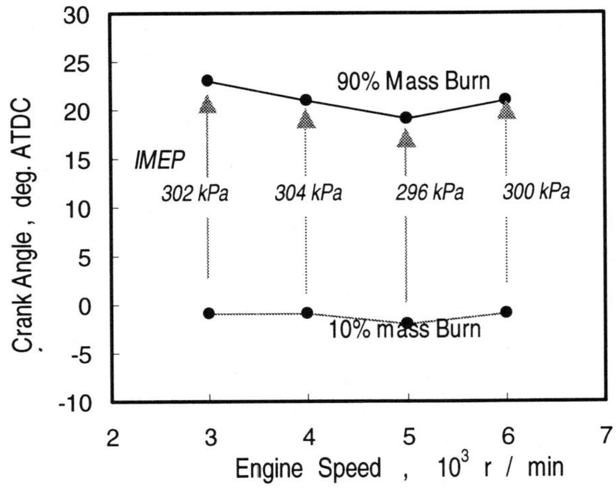

Figure 11. Combustion duration on the crank angle against the engine speed.

ADDITION OF AN AIR PUMP

The air pump was applied intending to a reliable PDI in the higher engine speed operation, however at the same time, it gave the engine a means to control the air amount for the PDI event. Especially at the AR combustion zone, its fine control became a subject. The additional air pump would reduce or enlarge the AR combustion zone? How does it affect the AR combustion performance such as the fuel consumption and the exhaust emissions? Considerable factors for the questions would be the pumped air temperature and the additional friction. The most optimistic estimation of ours was as follows. The outlet temperature of the pumped air reaches 400 K to 500 K, while the fresh air temperature from the scavenging ports does not exceed 400 K. Then the in-cylinder gas temperature after the gas exchange process would rise. It should reduce the AR combustion lower load limit. And, again, additional friction due to the air pump should promote it.

PDI air effect on the scavenging

The lower load limit of the AR combustion observed in the engine speed zone lower than 5,000 r/min is shown in figure 6. It is a limit of the heat balance maintaining the AR combustion. The heated PDI air would enhance the in-cylinder gas temperature? The delivery ratio required for the lower limit is generally less than 20%, while the Las could vary from 1% to 20%. Thus it will not be negligible the influence of the additional air for the PDI on the total fresh air in the cylinder. Consequently the summation of the delivery ratio and the Las should be regarded as the total fresh air supplied to the engine. Again, what complicates the matters is the additional air affects the engine's original scavenging characteristic. In the experimental engine the PDI event takes place just after the scavenging process while the exhaust port slightly opens. Therefore it is like a double scavenging. At the first scavenging process, a part of the intake air should short-circuit as the manner of conventional engines. However, at the second scavenging process, all gas would remain in the cylinder as long as the PDI perfectly performs the stratified scavenge. Figure 12 explains the mechanism of this double scavenging process by using the scavenging characteristic diagram and the exhaust gas analyzing results. The heavy solid line represents the experimental engine's original scavenging characteristic.

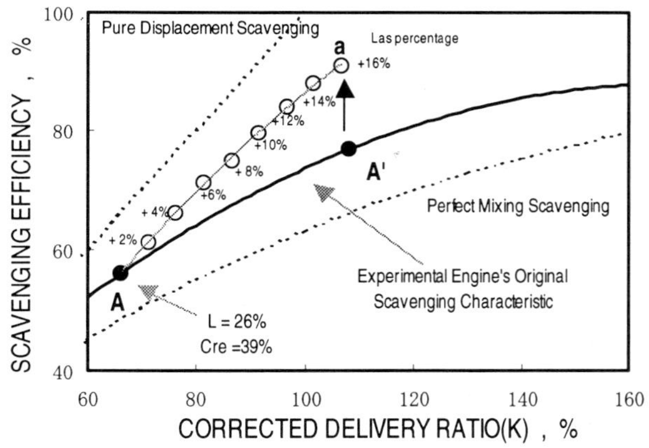

Figure 12. Effect of the Las on the scavenge.

Citing an instance of the black point A at the delivery ratio (L) of 26% and the relative charge (C_{rel}) of 39%, which is a typical part load in the AR zone, the original scavenging efficiency is 56% at a state of no additional air. As the Las increased, the scavenging efficiency increased on the line as the white points indicate. This line was nearly parallel to the line of the pure displacement scavenging, which meant that no additional gas short-circuited. If the same amount of the air were given from the intake port the scavenging efficiency would follow the original scavenging characteristic line to the point A', where is the L of 42% and the η_s of 77%. In the case of the PDI, an L of 26% plus an Las of 16% makes the scavenging efficiency (η_s) of 91%, which is point a. This means that as the Las increases, the same IMEP could be generated with a smaller amount of total air due to the improvement in the scavenging efficiency.

<u>AR combustion lower limit</u>

The air pump effect on the AR combustion lower limit was experimented at a constant engine speed of 3,000 r/min. Various combinations of Las and the delivery ratio (L) that resulted in the lower limit of the AR combustion were tested as shown in figure 13. The stoichiometric in-cylinder air to fuel ratio condition was provided for the test. According to the increase of the Las, a required L decreased, while the total delivery ratio, namely L + Las, increased.

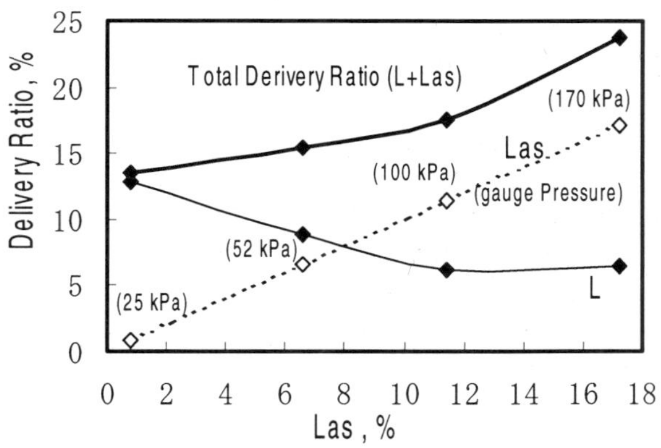

Figure 13. The delivery at the lower limit of the AR combustion in the various Las.

Figure 14 represents a series of these variations from the frictional mean effective pressure (FMEP) aspect and mean effective pressure (MEP) aspect. As Las increases, the FMEP of the engine slightly decreases according to the decrease of L, while the FMEP of the air pump increases according to the increase of the Las. The increase of the pumping loss of the air pump is larger than that of the engine due to the higher outlet pressure. Thus, total FMEP increases as the total delivery ratio increase. Meanwhile, the upper graph of the figure 14 shows the IMEP and BMEP at the AR combustion lower limits. The data shows that the IMEP of the lower limit has increased as the Las increased, being contrary to our primary optimistic prediction. The increase in the Las improves the scavenging efficiency as it has been formerly explained.

However, at the lower limits, the exhaust valve is at the fully closed position, therefore it does not have a further capability of controlling the scavenging efficiency. That would be the mechanism. This tendency shown in the IMEP characteristic is alleviated in the BMEP characteristic due to the FMEP characteristic. The addition of the air pump, ultimately, has not lowered the lower limit of the AR combustion.

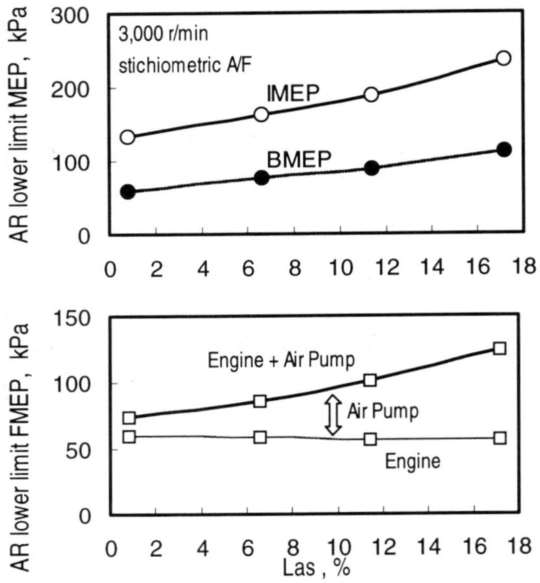

Figure 14. IMEP, BMEP and FMEP at the lower limit of the AR combustion at the various Las.

Effect of the PDI air on the AR combustion

Another concern of ours was how the Las would effect the AR combustion performance. As an instance, this paper mentions a typical AR combustion point at 4,000 r/min, an IMEP of 300 kPa. First of all, it was examined the combinations of the delivery ratio and the exhaust valve opening against the Las variations, which generated the AR combustion at a constant IMEP. As the Las increases, total delivery ratio for a constant IMEP gradually decreases and the exhaust valve opening required for the AR combustion decreases, shown in figure 15. This indicates that the exhaust valve compensates for the scavenging efficiency at a constant rate, in order to make the auto-ignition occur at a constant timing. Figure 16 shows the variations of the BMEP and the specific fuel consumption (SFC) against the Las variations. A stoichiometric air to fuel ratio is provided for the in-cylinder mixture throughout the experiment. The indicated specific fuel consumption (ISFC) remains virtually constant against the Las variations, while the brake specific fuel consumption (BSFC) deteriorates markedly as the Las increases. Thus, this deterioration of the BSFC should be due to the increase in the frictional loss, as it is substantiated by the deterioration of the BMEP. If the source of this friction rise would be the pumping loss, it may be contradicting with the total delivery ratio data in figure 15, which shows a decrease. The adiabatic efficiency of a reciprocating air pump deteriorates as the outlet pressure rises, which would be the cause of this friction rise.

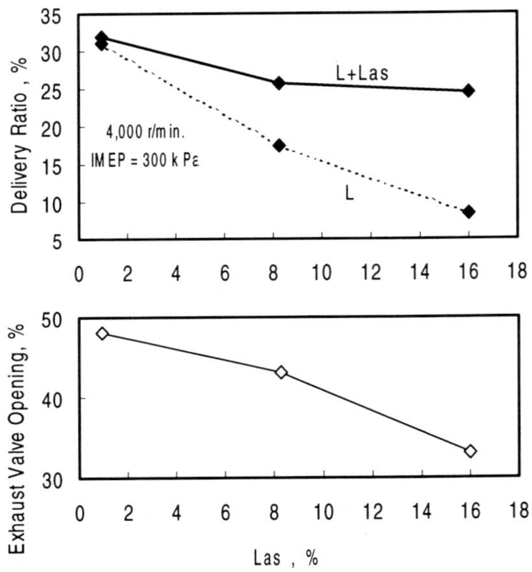

Figure 15. The delivery ratio and the exhaust valve opening for the AR combustion of a constant IMEP against the Las variations.

Meanwhile, emissions' data in figure 17 shows an interesting aspect. As the Las increases, the brake specific hydrocarbon (BSHC) increases in the same manner as the BSFC, however the brake specific nitrogen oxides (BSNOx) remarkably decrease. And this suggests a difference in the combustion processes.

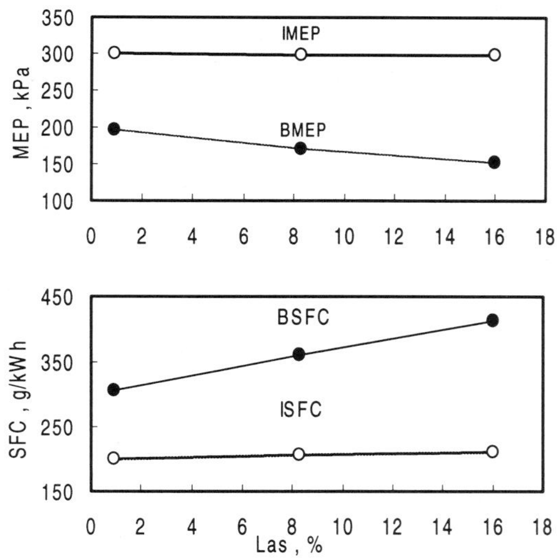

Figure 16. Variations of the BMEP and the SFC against the Las variations.

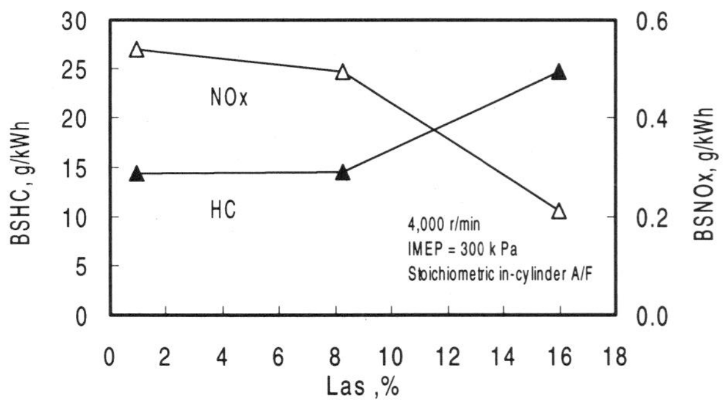

Figure 17. Variations of BSHC and BSNOx against the Las variation.

Intending to explicate the difference in these AR combustion processes, the indicator diagrams were analyzed. The AR combustion at the same IMEP and the same engine speed, with different Las, the cyclic variations in the indicator diagrams had apparently no difference. Figure 18 compares the heat release rate curves in the combustion with the Las of 1% and Las of 16%, in which an evident difference is observed in the curves. The heat release occurs more abruptly 16% case than 1% case. Figure 19 shows the heat release comparison. The onset timing of the heat release was however the same, combustion speed was evidently faster in the AR combustion with a larger Las. This data appears to make a difference in the scavenging efficiency, however in this case both scavenging efficiency, i.e. the fresh gas overall rate to the total gas in cylinder, was completely the same. If it were so, would it be a difference in the mixture formation owing to a large amount of leaner mixture injection and a small amount of richer mixture injection?

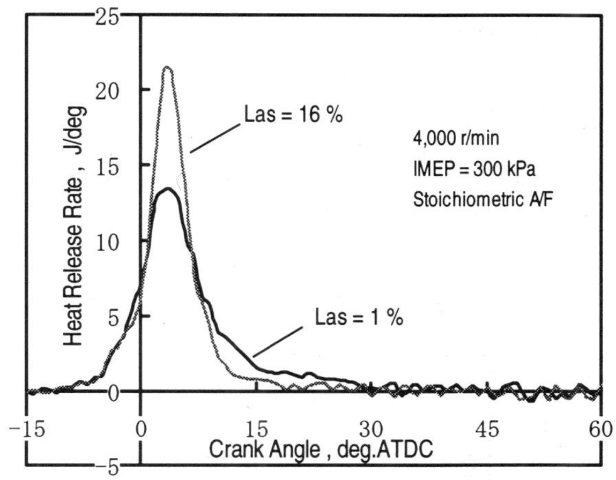

Figure 18. The heat release rate curves with the different Las.

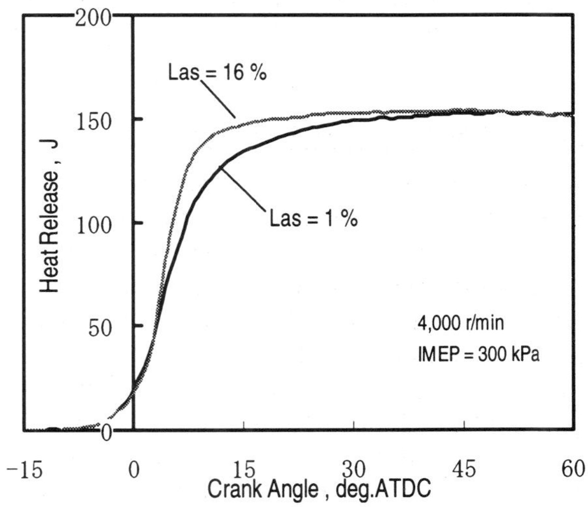

Figure 19. The heat release curves with the different Las.

DISCUSSION

This paper has mentioned fragmentary results of the experiments in which we took a primitive interest during the development of a high-speed high-power pneumatic direct injection two-stroke engine. Hence, it may lack a coherent explanation and an assertion as a rigorous technical paper. However if these fragments of the auto-ignition observations would become a hint of the understanding and the discussion, it would be a pleasure of ours. In the previous studies, we have focused a typical state of the gas exchange in the two-stroke engines. In order to put the auto-ignition to practical use, we have proposed a method to control its onset timing. And the method was demonstrated in several motorcycle engines. However once the onset timing was controlled, the combustion process was as it would be. If the auto-ignition were a manner of a compression ignition, what does it mean duration of the combustion? This time, the experimental data have revealed that the auto-ignition combustion speed could vary, depending on the concentration of the in-cylinder inert gases, i.e. scavenging efficiency, depending on the engine speed, etc. How would these factors affect the combustion process? They are all convenient features for the two-stroke engine application, however. According to our high-speed CCD camera observation, the auto-ignition primarily occurred from several kernels. Then the kernels grew to fill the space between the kernels and completed its combustion process. The kernels growing process also appeared to be local flame propagation. This mechanism of the combustion speed could be a key for the further applications of the auto-ignition to the internal combustion engines. In combination with the air pump for the PDI, unfortunately, it has not shown any extraordinary result or possibility of the AR combustion. The relation between the PDI pressure and the pumping loss would be a subject. However it was very interesting that the air amount difference for the PDI affected the AR combustion speed. From the viewpoint of lowering the lower limit of the AR combustion, it would be necessary for the auto-ignition studies to investigate farther from the formation aspect in the mixture and the inert gas. The AR combustion has been ascertained in a high-speed, high-power PDI engine and at the same time it left us several subjects as well.

SUMMARY

This study gave promise of application of the activated radical (AR) combustion concept to the high-speed high-power pneumatic direct injection (PDI) engine that generated a specific power output of 180 kW/ℓ at 11,000 r/min.

The experimental results by using the PDI-AR combustion engine, are summarized as follows,

1. The auto-ignition combustion speed varies depending on the operating conditions such as a load, a scavenging efficiency, an engine speed, and so on.
2. Duration between the 10% and the 90% of the mass burn timing in the AR combustion comes at the same crank timing regardlessly of the engine speed. Hence it follows that the AR combustion could be applicable to high-speed engines as well.
3. The air supply, which was a requisite to the PDI, affected the AR combustion both from the gas exchange aspect and from the frictional loss aspect.
4. The additional enthalpy of the air for the PDI was not sufficient for enhancing the onset of the auto-ignition. Conversely, as the air increased, the onset of it abated because of the improvement in the scavenging efficiency.
5. As the air supply increased, the total frictional loss of the engine increased due to the pumping loss of the air pump, however a BMEP at the AR combustion lower limit would not have come down because of the effect mentioned in 4.
6. The air supply for the PDI should be reduced in the AR combustion area for a lower BSFC or a lower BSHC.
7. It was revealed that the process of the AR combustion itself varied according to the quantity of the air for the PDI.

CONTACT

Yoichi ISHIBASHI ; Ph.D., Chief Engineer,
HONDA R&D Co.,Ltd. Asaka R&D center
3-15-1 Senzui Asaka-shi Saitama 351 JAPAN
e-mail : yoichi.ishibashi@mail.a.rd.honda.co.jp
Telephone : +81.48.462.3285, Telefax : +81.48.462.2954

REFERENCES

[1] Jean C. Dabadie, Pierre L. Duret and Stéphane Venturi, "SCIP: a New Simplified Camless IAPAC Direct Injection for Low Emission Small Two-Stroke Engines", SAE 972078, SETC'97, Yokohama, Japan

[2] Mark Archer and Greg Bell, "Synerject Injection Systems – An Emissions Solution for both 2- and 4-stroke Small Vehicle Engines", SIAT'2001, Pune, India

[3] Marco Nuti, Maurizio Marcacci, Roberto Pardini and David Caponi, "FAST Injection System: A Very Simple Way to Lean Combustion in SI Engines", SAE 972073, SETC'97, Yokohama, Japan

[4] Koji Morikawa, Hideharu Takimoto, Yasuo Moriyoshi and Yasuo Imai, "Design and Development of Newly Proposed Reverse Uniflow-type Two-Stroke Gasoline DI Engine", SAE 2001-01-1095, SAE annual meeting, Detroit, USA

[5] Yoichi Ishibashi, Masahiro Asai and Kenji Nishida, "An Experimental Study of Stratified Scavenging Activated Radical Combustion Engine", SAE 972077, SETC'97, Yokohama, Japan

[6] Yoichi Ishibashi and Masahiro Asai, "A Low Pressure Pneumatic Direct Injection Two-Stroke Engine by Activated Radical Combustion Concept", SAE 980757, SAE annual meeting, Detroit, USA

[7] Yoïchi Ishibashi, "Basic Understanding of Activated Radical Combustion and Its Two-Stroke Engine Application and Benefits", SAE 2000-91-1836, SAE F&L Spring 2000, Paris, France

DEFINITIONS, ACRONYMS, ABBREVIATIONS

Definitions concerning Gas Exchanging State

Mf mass of fresh air trapped in the cylinder
Mr mass of residual gas trapped in the cylinder
Mg mass of total gas trapped in the cylinder
Ms mass of intake air for one cycle
Mh reference mass required to fill the displacement under the reference conditions.

Delivery Ratio	**L = Ms / Mh**
Trapping Efficiency	η_{tr} **= Mf / Ms**
Scavenging Efficiency	ηs **= Mf / Mg**
Charging Efficiency	ηc **= Mf / Mh = L x** η_{tr}
Relative Charge	C_{rel} **= Mg / Mh**
Corrected Delivery Ratio	**K = Ms / Mg = L /** C_{rel}

Symbols and Definitions

PDI: pneumatic direct injection

AR: activated radical

CRg : geometric compression ratio

Exhaust Opening Rate (EOR):
a rate of exhaust valve opening area against the valve fully open state.

Las: a rate of the air mass supplied for the PDI system against the reference mass required to fill the displacement under the reference conditions.

Total Delivery Ratio: L + Las

A New Generation of Engine Combustion Processes for the Future?
© P. Duret (Editor) and Editions Technip, Paris, 2001, pp. 153 - 162
27 rue Ginoux, 75015 Paris

RADICAL ACTIVATED COMBUSTION A NEW APPROACH FOR GASOLINE ENGINES

Peter Wolters[1], Wolfgang Salber[1], Jochen Dilthey[2]
1 FEV Motorentechnik GmbH 2 VKA Institute for Combustion Engines

INTRODUCTION

When Nicolaus August Otto was working on his pioneering invention of an internal combustion engine, he was struggling with an auto ignition phenomenon which, according to his reports, was about to destroy the engine structure. Since than, it was a central task of gasoline engine development to avoid auto ignition under any circumstances. This target has driven the development of engine technology as well as of fuel properties.

125 years later, engineers around the globe again are struggling, but now in order to get this phenomenon under control and make use of it. The idea is to run the engine with extremely diluted mixture which produces extremely low emissions, especially of nitrogen oxides, at part load operation. This could be the key to a significant improvement of part load fuel economy without drawbacks in emission behavior and exhaust aftertreatment complexity. Since such mixtures can not be ignited by spark plugs, controlled auto ignition (CAI) is on the road map of future gasoline engine technology.

Based on experimental results on a single-cylinder engine this paper presents a theory of controlled auto ignition called "Radical Activated Combustion". In a further part, the necessary design features of a commercialized combustion engine are addressed. Variable valve actuation as a means to control residual gas and variable compression ratio are in the focus of this technology review. Design solutions are presented and their potential contributions to a CAI engine concept are discussed.

RADICAL ACTIVATED COMBUSTION (RAC)

It has been observed very early that in certain operation areas piston controlled two-stroke engines keep running even after the ignition was shut off. This form of auto ignition in charge compressing two-stroke engines has been described as very smooth engine operation with low HC-emissions and largely reduced fuel consumption.

Within the scope of reducing the emissions of two-stroke engines this form of combustion has been closely investigated in the 70s, especially in Japan. Generally, RAC is characterized by very smooth engine operation and the reduction of emissions and fuel consumption.

This form of auto ignition has often been confused with surface ignition or knocking combustion. Contrary to these destructive forms of uncontrolled auto ignition, occurring at or near full load operation, RAC is only observed at low engine load.

Depending on the author or research organisation this combustion principle has been given various names, e.g.:

- ATAC, Active Thermo-Atmosphere Combustion
- PCCI, Premixed Charge Compression Ignition
- CAI, Controlled Auto Ignition
- AR, Activated Radical Combustion

- CIHC, Compression-Ignited Homogeneous Charge Combustion
- HCCI, Homogeneous-Charge Compression Ignition
- RAC, Radical Activated Combustion

Along with the worldwide search for new concepts to improve efficiency and exhaust emissions of gasoline engines, the principle of controlled auto ignition has recently aroused lively interest among engine developers.

THEORY OF RAC

Besides engine-related parameters the tendency of a gasoline engine to auto-ignite is depending strongly on the operation conditions of the engine. The pressure and temperature levels, at which RAC is occurring, are also present in the unburnt end gas of conventional SI engines during combustion but do not necessarily induce auto ignition. Next to the thermal conditions of the cylinder charge there must be other parameters affecting the process of RAC.

RAC is based on a two-stage ignition process. The first stage is taking place during the first phase of the compression stroke. At this pressure and temperature level cold-flame reactions are running in the air fuel mixture. These reactions sensitize the mixture, and they are obviously assisted by active intermediate products in the „active" residual gas from the last proceeding combustion.

These active intermediate products are not radicals in the chemical meaning but aldehydes and peroxides originating from the residual gas of the last combustion, whose dissociation starts the low-temperature flame during the compression, promoting the start of the hot flame.

According to /1/ the prereactions in the cylinder charge, leading to controlled auto ignition, are strongly depending on the amount, the composition and the thermal state of the residual gas inside the cylinder. Lingen /1/ claims for a two-stroke engine that the fraction of "active" residual gas, which has not been "stored" in the crankcase or exhaust port during one cycle, is decisive for the inclination of the engine to auto-ignite.

The second auto ignition stage is induced towards the end of the compression by the thermal dissociation of the aldehydes and peroxides during the first cold-flame reactions. This is corresponding to /2/, where varying amounts of formaldehyde were added to the intake air to trigger and to control the ignition and combustion on an RAC engine operated with Compressed Natural Gas (CNG).

Though /5/ refuses the influence of chemical intermediates in the exhaust gas on the ignition process and states that EGR has only thermal effects, other works show that there must be more to the influence of the EGR rate on the RAC process than a mere raise in in-cylinder mixture temperature.

Several works /4,7,8/ show that there is a strong influence of the EGR rate on the heat release during combustion and subsequently on the peak pressure position that cannot be explained by thermal influence of EGR alone.

/4/ presents several successful investigations to realize RAC on a 4-stroke engine with internal EGR via fully variable valve actuation. It was possible to control the point of auto ignition in a wide area by the amount of EGR. Higher EGR rates caused earlier auto ignition, shifting the combustion peak towards TDC.

Lingen /1/ has shown in his work, that if the fraction of "inactive" residual gas, not originating from the very last combustion, is exceeding a certain amount, the engine will no longer auto-ignite, almost regardless of the thermal conditions during compression.

Fully variable valve actuation, permitting complete control of internal EGR seems to be a very promising concept to realize an engine, that can be operated with RAC in most of its operation range.

ENGINE CONCEPT

Application of RAC to four-stroke engines is often described as requiring either extreme compression ratios like diesel engines, feeding of an additional low-octane fuel or drastic heating of the intake air. These steps taken to provoke controlled auto ignition on gasoline engines put a question mark to the possible use of the RAC combustion process in vehicle engines.

The theory shows that a control of RAC requires certain tools to influence the important boundary conditions. From our point of view these are the possibility to vary the engine's compression ratio as well as a variable valve actuation device, which enables appropriate EGR strategies. Different to intake air preheating, these concepts allow transient operation under typical vehicle related conditions.

VARIABLE COMPRESSION RATIO

The patent literature offers manifold possibilities for the realization of a variable compression ratio. A select choice of these possibilities is shown in Figure 1.

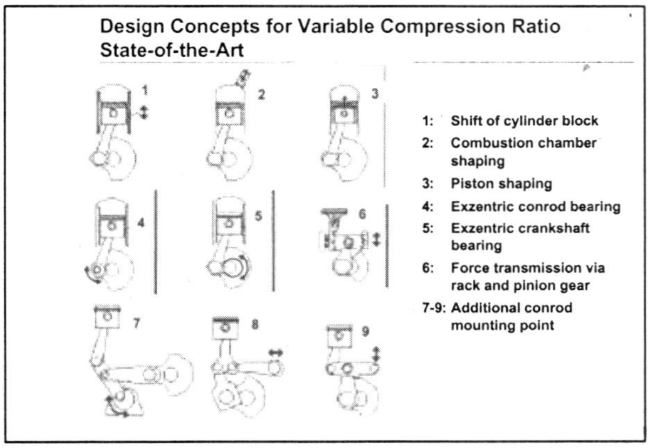

Figure 1: Concepts for Variable Compression Ratio

One obvious possibility is a variable combustion chamber volume through a secondary piston in the cylinder head. This design has been implemented for 2-valve cylinder heads /12/. The implementation becomes more difficult for 4-valve cylinder heads. Moreover, this solutions is disadvantageous in context with a sophisticated combustion system, which requires a certain combustion chamber shape.

This problem can be avoided by a piston with variable compression height or a connecting rod with variable length respectively. Among other reasons, the significant increase of oscillating masses has put the piston solution out of interest. In the mid-nineties FEV realized a variable compression concept with the aid of a split connecting rod and an additional articulated rod /13/, which is shown in Figure 2. In the meantime, this principle has been implemented in several engines and it has also been operated in vehicles. The main drawback of this solution is related to the fact that it is far away from existing bottom end designs.

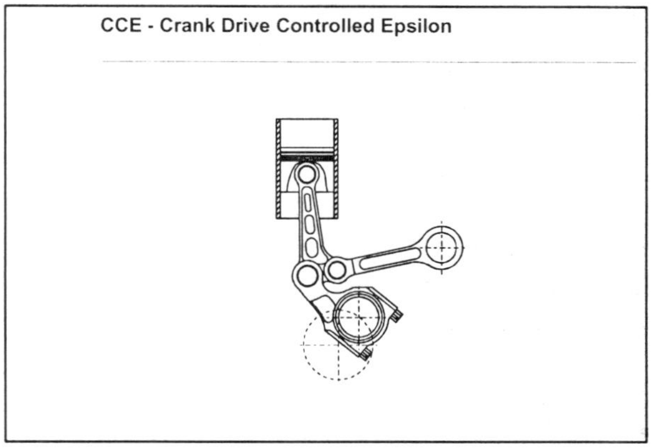

Figure 2: Crank Drive Controlled Epsilon

The basic principle of repositioning the cylinder or the cylinder head has recently been put into practice by Saab in its SVC engine /14/. The engine block is more or less "folded up" and thus the compression volume is changed. Again, this solution requires a complete new engine architecture and it seems impossible to integrate it into existing engine families.

A new trend setting technology is currently developed at FEV. The principle is depicted in Figure 3. The crankshaft turns in eccentric bearings which allows a controlled change of its position (CCP - Controlled Crankshaft Position). Through rotation of the eccentrics the vertical position of the bearing is changed and thus the top and bottom dead center of the piston are evenly shifted. The positioning of the crankshaft axis has to be balanced at timing and flywheel drive by special devices, which will be presented below.

Compared to the other approaches the CCP solution presents itself rather favorably. A convenient realization of the design effort is crucial for its implementation. The following factors are decisive for the choice of the CCP solution:

- no influence on the combustion chamber geometry
- small force demand of the positioning device
- reasonable design effort
- no change of the inertia forces
- good adjustability
- compatible to most production base engines
- high investments in new production units should not be necessary, conventional facilities should remain usable
- main dimensions may remain essentially unchanged, almost unchanged package conditions are about to apply

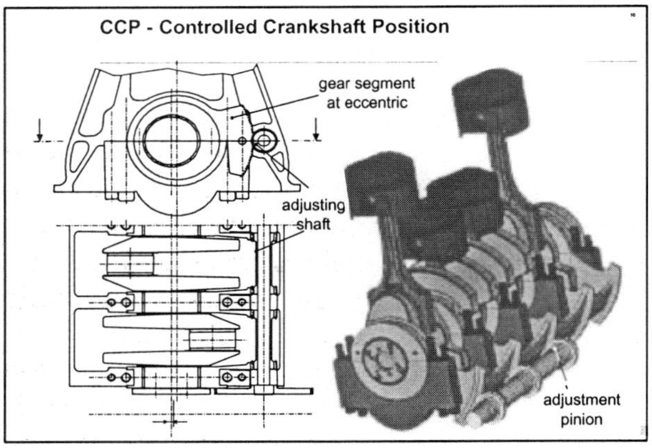

Figure 3: Controlled Crankshaft Position

As already mentioned before, the relative movement of the crankshaft requires a special connection of the flywheel and of the timing drive pulley. An obvious connection which can be used on both sides is an eccentric gear with a toothed annulus and an externally toothed spur. Figure 4 shows the design of the flywheel side which is currently used on a 4-cylinder test engine. The annulus of the eccentric gear is tightly connected to the crankshaft. In its further function the annulus represents the primary drive side of the flywheel, and the secondary drive side is represented by the spur wheel. The spur wheel bearing is arranged stationary in the flywheel case and connected to the clutch (which is not shown here).

The presented variable compression ratio device is currently investigated on a series production engine. This engine is a 1.8 l turbocharged engine. It was chosen with respect to its relatively high power density and modern engine design. The technical data which are specific for this concept are the following:

- max. eccentricity : 5.5 mm
- max. swiveling for CR-range : 63°
- compression ratio : 8 to 14

In order to realize this prototype engine the existing bulkheads were modified to allow the integration of the new main bearings. The series production connection rods were inserted without any alterations. The series production crankshaft could be adopted with minor modifications at the flywheel flange and the front crankshaft journal. The bearing positions for the main bearings as well as the connecting rod bearings were retained without modifications. This implies no design restraint for the layout of the bearing positions in the eccentrics. All over, this test engine concepts proves the feasibility of the CCP concept for mass production at reasonable costs.

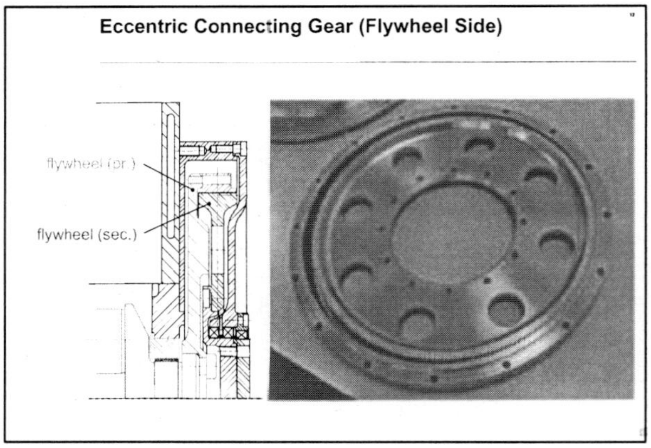

Figure 4: Eccentric Connecting Gear

The eccentrics were designed according to the series engine layout and designed in correspondence with the mechanical requirements. Series bearing shells were used. The bearing pre-stresses prescribed by the plain bearing manufacturer were realized by a corresponding bolting of the eccentrics. To avoid twisting of the eccentrics towards each other and hence the occurrence of alignment errors, the sector gears required for the positioning were bolted together. Thus, the eccentrics were rigidly connected through the link plates.

VARIABLE VALVE ACTUATION

Throttle-free load control represents a promising concept for achieving a clear reduction in the gas exchange losses in a gasoline engine. One of the most promising methods to reduce fuel consumption is to use unthrottled engine operation, where load control occurs by means of variable valve timing with an **E**lectro-**M**echanical **V**alve **T**rain (EMVT) system. In this concept the engine load is not controlled by the density of the air mass, but rather, by the opening duration of the intake valves. This method allows a reduction in fuel consumption while operating under a stoichiometric air-fuel-ratio and preserves the ability to use conventional exhaust gas aftertreatment technology with a 3-way catalyst. Results in the **N**ew **E**uropean **D**riving **C**ycle (NEDC) have confirmed a reduction in fuel consumption of more than 15% while meeting EURO IV emission limits /9/.

Engine Load Control

One of the primary reasons for developing EMVT is the benefit of controlling engine load without incurring throttling losses. For load control purposes, "Early Intake Closure (EIC)" and “Late Intake Closure (LIC)” are the strategies to control the quantity of aspirated mixture (Figure 5).

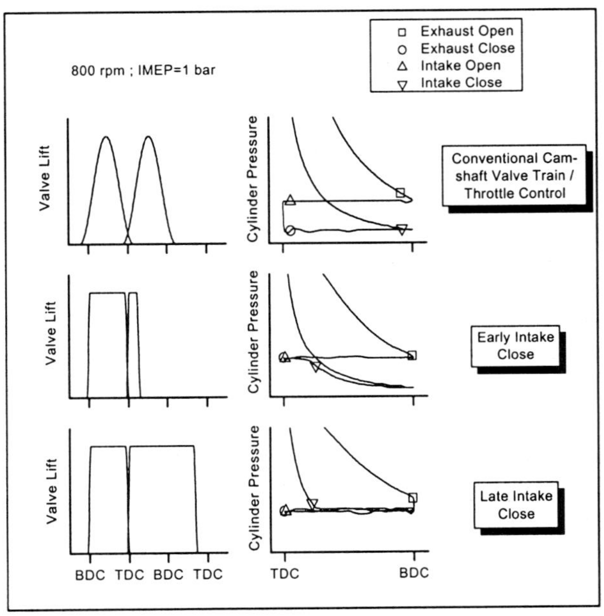

Figure 5: Valve Lift and Cylinder Pressure for Different Load Control Strategies

With the EIC load control strategy, the intake valves are closed during the intake stroke before BDC at the point where the desired fresh mixture mass is in the cylinder. With the LIC load control strategy, the surplus fresh mixture mass is pushed back out of the cylinder on the compression stroke through the opened intake valve. In comparison with the "Early Intake Closure" strategy, this leads to a light increase in pumping losses, especially at part load. Furthermore, the comparably lower efficiency with the LIC strategy (compared to EIC) can be explained by the larger standard deviation of the indicated mean effective pressure (IMEP) that originates from changing mixture qualities in the intake port. With unthrottled load control, the "Early Intake Closure" strategy is preferred at part load and low engine speeds.

Control of Load Composition

In addition to load control, the control of charge composition is a key starting point with regard to efficiency enhancements, emissions reductions and especially for the auto ignition combustion process. The variable valve control system makes it possible to adapt the residual gas quantity (essentially determined by the "intake open" and "outlet close" valve timings) to the particular operating point and state of the engine. The exhaust mass, which is recirculated, is also influenced by the duration and the phasing of the valve overlap. Apart from accurate control of the cylinder charge, also a high internal temperature inside the cylinder can be achieved with the electromechanical valve train for the RAC combustion process depending on the recirculation strategy for the residual gas control.

As shown in Figure 6, for an engine with variable intake and exhaust valve control three different procedures are possible to realize internal EGR.

- EGR through the intake port
- EGR through the exhaust port
- EGR within the combustion chamber

In the case of EGR through the intake port, the residual gas amount is determined by the timings of "Intake Open" and "Exhaust Close". After opening the intake valves, a portion of the exhaust gas is pushed through the intake valve opening cross section into the intake port during the exhaust stroke. Therefore, for unthrottled load control, a large valve overlap leads to a more significant dependence of load exchange on intake port and the exhaust port dynamics. However, to avoid superfluous pumping losses, the valve overlap should not be too small. The recirculation of hot exhaust gas into the intake port leads to higher mixture temperature in the intake port and consequently, to improved mixture preparation.

For EGR through the exhaust port, the residual gas amount is primarily determined by the "Exhaust Close" timing. Until closure of the exhaust valves after TDC, exhaust gas flows back from the exhaust port into the cylinder during the early intake stroke. The valve overlap length is optimized by the need to avoid pumping losses as described above for intake port EGR. This strategy leads to higher temperatures in the cylinder and a good homogeneity of charge because of the parallel aspirated fresh air and residual gas during the intake stroke. So this strategy is more suitable for the auto-ignited combustion compared to the exhaust gas recirculation through the intake port.

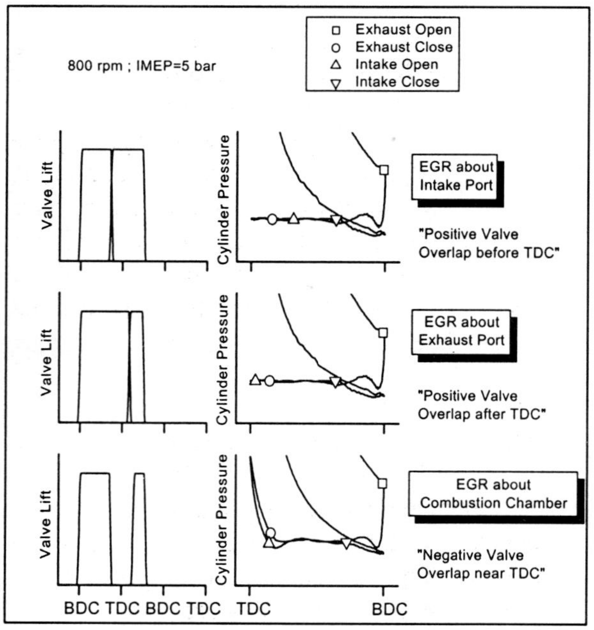

Figure 6: Residual Gas Control for an Engine with Variable Valve Timing

In contrast to the procedures described above, internal EGR within the combustion chamber requires a negative valve overlap. The exhaust valves are closed before TDC so that exhaust gas remains in the cylinder. The residual gas is compressed and expanded again after TDC until the intake valves open. Due to the minimization of the pumping losses, the intake valves are opened when the pressures in the cylinder and the intake port are equivalent. For this type of exhaust gas recirculation the in-cylinder temperature is on the highest level.

Control of In-Cylinder Gas Motion

During the gas exchange process the state of flow of the aspirated charge is crucial for its result. The prime objectives in generating charge movement are the improvement of the mixture formation and an increase in the internal cylinder flow to improve homogenization of the stoichiometric mixture and to accelerate combustion.

One strategy to increase charge motion by means of the variable valve control is using a gas exchange with negative valve overlap. To increase the in-cylinder gas motion, it is necessary to open the intake valves after TDC. With negative valve overlap and late opening of the intake valves after the TDC, a low pressure level originates from the downward movement of piston in the cylinder. The magnitude of the flow velocity is determined by the pressure difference between intake port and cylinder when the intake valves open. The critical pressure ratio, which causes the mixture to flow into the cylinder at sonic speed, is indicated when the pressure in the intake port is 1 bar and the cylinder pressure is less than approximately 0.56 bar.

Late Intake Open ("LIO") gas exchange strategies are primarily utilized to improve combustion behavior at low engine load. Figure 7 shows two possible concepts, which were optimized for idle operation. The exhaust valve is closed in the area of TDC to minimize the residual gas fraction for the idle point that is displayed. With this gas exchange strategy, exhaust gas recirculation is possible by closing the exhaust valve before or after TDC. Accordingly, to achieve the same flow rate, the intake valve must be opened later.

In the upper part of the Figure 7 the LIO gas exchange concept is shown with a "Late Intake Close" strategy. The indicated p-V-diagram clarifies that the pumping losses increase further when the intake-valve is opened late. The pumping losses only decrease when the intake valve is opened very late after BDC in the compression stroke because of the near congruence of the expansion and compression line. The later intake of the mixture increases the in-cylinder gas motion at ignition and during combustion.

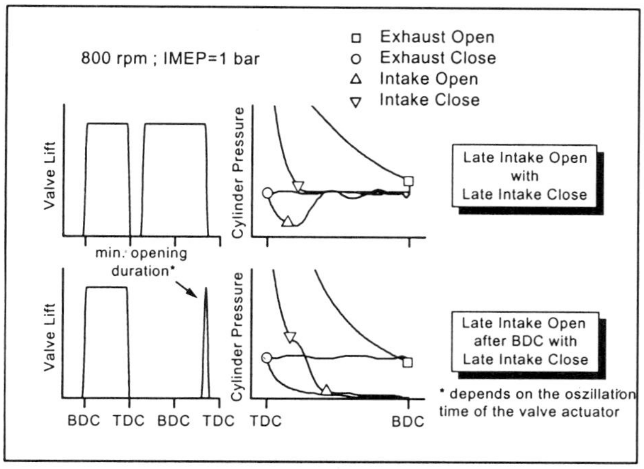

Figure 7: Gas Exchange Strategy "Late Intake Open" for an EMVT Engine

Opening the intake valve during the compression stroke with minimal valve open duration (bottom part of Figure 7) will generate the maximum in-cylinder gas motion during combustion. Since the cylinder pressure is greater than the ambient pressure level while the intake valve is closed, a back flow of the mixture takes place during the minimal valve open duration.

The LIO strategy is used especially during cold start and catalyst heating period. Increasing the charge motion during this phase by opening the inlet valve late with a negative valve overlap produces (due to the high flow velocity in the valve slit from as early as the first working cycle) a very stable combustion and a general reduction in hydrocarbon emissions, particularly in the main conversion and post-combustion phase /10/. The high intake velocity due to the late opening inlet valve also produces a significant reduction in the fuel film on the intake port wall. This permits improved metering accuracy and, in addition, an improvement in the transient behavior under no-load conditions.

For the RAC the LIO strategy is useful together with the EGR through the exhaust port or within the combustion chamber. Because of the higher in-cylinder gas movement the homogenization of the charge is improved significantly.

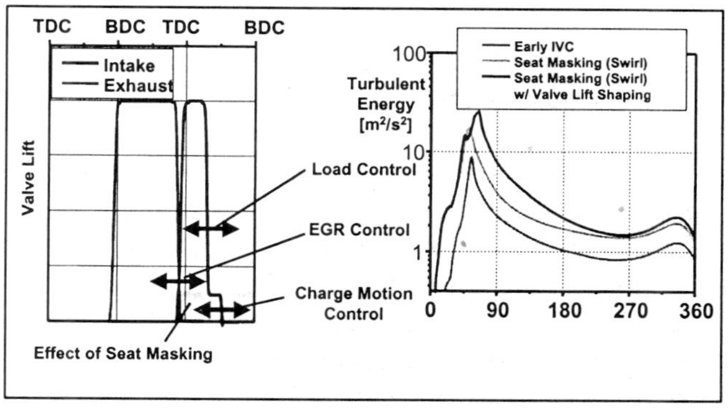

Figure 8: Combustion Process Optimization by Means of Valve Lift Curve Shaping

Further potential to improve in-cylinder gas movement for a variable valve train results from a reduction of valve lift and from intake valve deactivation. Current electromechanical valve control systems are based on a control concept for the valve actuator that allows adaptation of the valve lift and the valve movement near the seating positions /11/. The primary motivation in adapting this type of control is reducing the seating velocity of the armature plate and the valve to minimize valve train noise and improve durability. Reproducible seating velocities below 0.05 m/s are currently achieved. Today's applications of the electromechanical valve train also demonstrate that it is possible to shape the valve lift curve with the aid of this advanced valve actuator control (Figure 8).

As a result, it is possible to achieve a high flow velocity in the valve slit via the gradient of the valve lift during opening and closing and to optimize the homogenization and the combustion. To improve the effect of the valve lift curve shaping also a masked valve seat is used for this strategy.

Valve timing "Exhaust Open"

The exhaust open valve timing influences the process efficiency, the exhaust gas temperature level and the secondary reactions of the hydrocarbon emissions in the cylinder and in the exhaust system and so again the exhaust gas temperature. Therefore, a careful adaptation of the exhaust valve open timing can be used to influence the residual gas temperature for RAC. The effect of a late opening of the exhaust valve is a reduced backflow of exhaust gases from the exhaust port back into the cylinder at part load. This leads to a higher temperature in the cylinder, because the dilution of hot gases with colder gases from the exhaust port is avoided. Also the longer high temperature period before opening the exhaust valve promotes the secondary reactions in the cylinder and, thus, a higher temperature level in the cylinder.

Strategy for RAC with Electromechanical Variable Valve Train

Investigations with RAC in an engine equipped with an electromechanical valve train show that the internal residual gas control and late opening of the outlet valve present conditions that favor auto ignition towards the end of the compression phase (Figure 9).

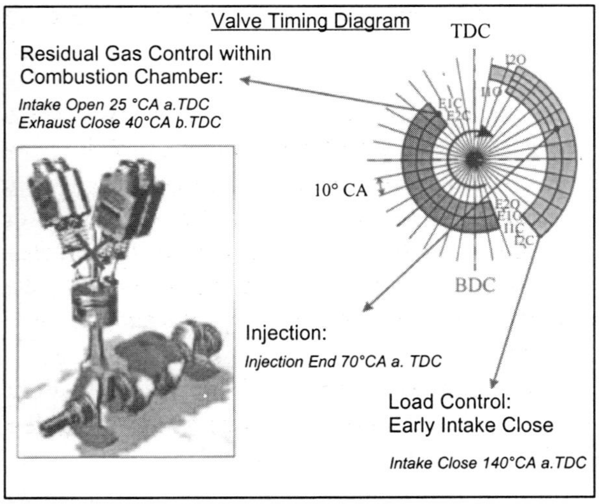

Figure 9: Load Exchange Control for Radical Activated Combustion with an Electromechanical Valve Train

In this regard, it is advantageous if the residual gas remains at a high temperature during the gas exchange. This is achieved through a combustion chamber residual gas recirculation concept, which does not exhaust the residual quantity of gas at first. Together with a LIO strategy an excellent homogenization is achieved. Also the strategy with EGR through the exhaust port leads to a high temperature in the cylinder. Because of the parallel aspiration of residual gas and fresh air a good homogenization is reached also without LIO.

Even a two-stroke mode can be realized with the electromechanical valve train. As the results in /1/ teach, the shortened time interval between successive combustion events favors auto ignition. Due to the cycle-synchronous operating mode of EMVT the engine mode can be changed from cycle to cycle. This enables an effective use of different combustion modes at transient operation .

INFLUENCE OF EGR COMPOSITION

As already mentioned before, the recirculated exhaust gas has not only purely thermal effects on the fresh charge. Moreover, there is also an influence of the composition and condition of this residual gas, that can be controlled with gas exchange strategies offered by fully variable valve actuation. This influenc of the EGR composition was investigated on a two-stroke engine /1/. The engine was equipped with adjustable overflow throttles in the intake and exhaust ports, so the exhaust gas composition can be varied by opening and closing these throttles.

When the intake port throttle is opened, a certain amount of residual gas can enter into the crankcase through the transfer ports. This exhaust gas is mixed with fresh air-fuel mixture. A part of this gas composition enters the combustion chamber immediately during the scavenging phase. The remaining partition stays within the crankcase until the next cycle.

When the transfer port throttle is closed, there is less back flow of residual gas into the crank case. Now the residual gas fraction must be adjusted by the exhaust port throttle. Thus, the residual gas originates more from the last combustion cycle and is less stored in the crankcase for longer duration. Although the calculated overall EGR ratio is the same, the state of it is significantly different. The test results show that in spite of almost identical temperature levels during the compression the combustion turns up earlier and proceeds faster during the operation with the restricted admission port and thereby the higher amount of fresh residual gas (Figure 10).

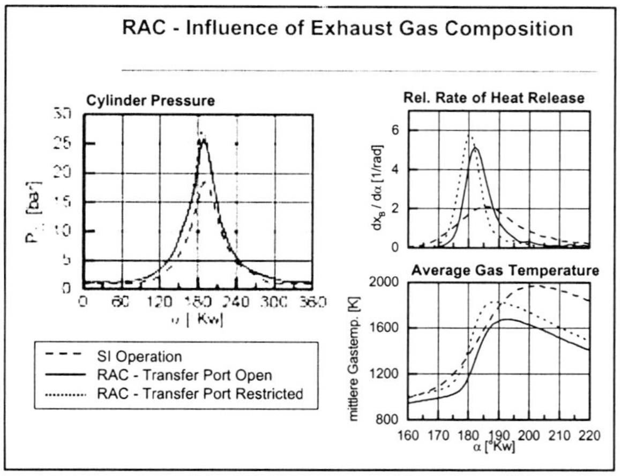

Figure 10: Combustion with RAC

Figure 11 shows the EGR ratios for these tests. At standard SI operation the major part of the residual gas comes from the crankcase, where it has been stored during the last cycles and can be considered inert. Under these conditions no RAC is observed.

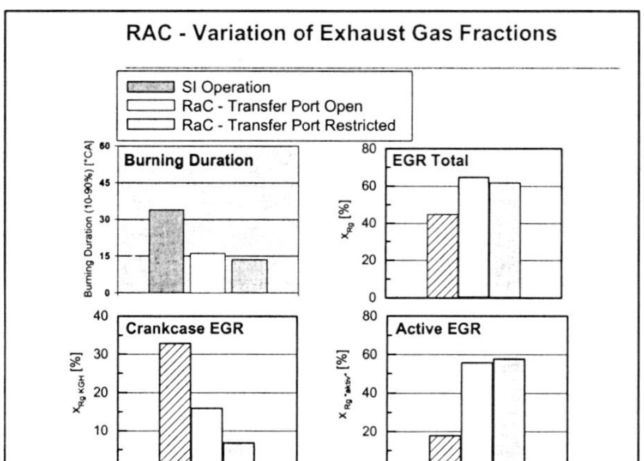

Figure 11: Exhaust Gas Fractions at RAC

When the total EGR ratio is increased, RAC occurs. With throttled transfer ports the fraction of fresh residual gas increases and the fraction of crankcase-stored residual gas is accordingly reduced compared to the operation with unrestricted transfer ports. This leads to a shorter burning duration.

To adopt the principle of RAC on a four-stroke engine, it must be ensured, that the recirculated exhaust gas does not become inert but comes into contact with the fresh charge quickly. A very promising concept to realize this is fully variable valve actuation. First the exhaust gas can be kept hot in the cylinder by late opening of the exhaust valves. Then there are different ways to thoroughly mix fresh and exhaust gas by applying various valve timing strategies, as described in /4/.

Additionally cycle consistent switching between different control strategies can be easily realized. This renders the possibility to change between SI and RAC control strategies transiently in different map areas.

In combination with a variable control of compression ratio RAC can be applied in a valuable area of the part load engine map.

SUMMARY

Controlled auto ignition might represent a breakthrough in engine technology. It combines high efficiency with ultra low raw emissions. The results derived from research engines are promising and justify expansive research in this new type of combustion.

Engine tests, which were carried out on a two-stroke engine, showed a significant influence of the state of residual gas. The results prove that not only the total EGR ratio is decisive for the occurrence of auto ignition. Beside this, the history of recirculated gas starting from its production during the combustion process and ending with auto ignition plays an important role. This also explains experiences, which were reported in other publications and which showed differences due to the realization of exhaust gas recirculation.

For auto ignition it is favorable to keep the recirculated gas on an "active" state. That is the reason, why we chose "Radical Activated Combustion" (RAC) as an acronym.

Therefore, EGR via the intake ports cannot be as beneficial as a solution, where the exhaust gas is either catched within the cylinder or is sucked out of the exhaust ports.

Since these preferred strategies require unconventional valve timings, variable valve actuation is a key element to utilize the benefits of RAC under vehicle related operating conditions. Especially electromechanical valve actuation has a maximum potential of flexibility. FEV has developed EMVT technology to a degree, which makes it a first choice for RAC applications.

As to the variable control of compression ratio the eccentric crankshaft positioning concept represents a design solution, which has almost no negative effects on engine function. Moreover, an eccentric positioning of the crankshaft can take place with relatively low actuating forces. The conversion of the crankshaft movement into a rotation with stationary axis at the timing drive and flywheel end is crucial for the limitation of the design effort. The solutions developed in this field were able to prove their feasibility on the engine test stand and represent an overall concept which is suitable for series production.

However, the application of controlled auto ignition will demand a lot of further basic research. The main target will be to understand the physical and chemical aspects of the process in order to minimize the necessary complexity of control.

REFERENCES

1. A. Lingen
   Untersuchungen der Radikalen-aktivierten Verbrennung als Brennverfahren für gemischverdichtende Motoren
   Dissertation, RWTH Aachen, 2000

2. Y. Ohta, M. Furutani, M. Kojima, M. Kono
   Premixed-Compression-Ignition Natural-Gas Engine. FISITA 2000

3. Y. Ohta, M. Furutani
   Identification of Cool and Blue Flames in Compression Ignition. Aarchivum Combustionis Vol.11 (1991), Polish Academy of Sciences

4. D. Law, D. Kemp, J. Allen, G. Kirkpatrick, T. Copland
   Controlled Combustion in an IC-Engine with a Fully Variable Valve Train. SAE 2000-01-0251

5. P. M. Najt, D. E. Foster
   Compression-Ignited Homogeneous Charge Combustion. SAE 830264

6. R. H. Thring
   Homogeneous-Charge Compression-Ignition (HCCI) Engines. SAE 892068

7. Lavy et al. (14 Authors)
   Innovative Ultra Low NOx controlled Auto Ignition Combustion Process for Gasoline Engines: the 4th SPACE Project.
   SAE 2000-01-1837

8. N. Lida, T. Igarashi
   Auto-Ignition and Combustion of N-Butane and DNE/Air Mixtures in a Homgeneous Charge Compression Ignition Engine.
   SAE 2001-01-1832

9. H. Baumgarten, H. Kemper, M. Pischinger, W. Salber, F. van der Staay
   Low Fuel Consumption and Low Emissions – Electromechanical Valve Train in Vehicle Operation
   Seoul 2000 FISITA World Automotive Congress

10. T. Esch, H. Kemper, W. Salber, F. van der Staay
    Der elektromechanische Ventiltrieb
    Systembaustein für zukünftige Antriebskonzepte
    MTZ 12/2000 und MTZ 1/2001

11. S. Pischinger, W. Salber
    Möglichkeiten zur Verbesserung des Kaltstart-, Warmlauf- und Instationärverhaltens mittels variabler Ventilsteuerzeiten
    20. Internationales Wiener Motorensymposium, VDI Fortschrittberichte Reihe 12 Nr. 376, 1999

12. P. Walzer, P. Adamis, H. Heinrich, V. Schumacher
    Variable Steuerzeiten und variable Verdichtung beim Ottomotor. MTZ 47 (1986) 1

13. C. Bollig, K. Habermann, H. Marckwardt, K. I. Yapici. Kurbeltrieb für variable Verdichtung.
    MTZ 58 (1997) 11

14. H. Drangel, L. Bergsten
    Der neue Saab SVC Motor – Ein Zusammenspiel zur Verbrauchsreduzierung von variabler Verdichtung, Hochaufladung und Downsizing
    9. Aachener Kolloquium Fahrzeug- und Motorentechnik, 4.-6.10.2000

WHAT ENGINE COMBUSTION PROCESSES FOR THE FUTURE?

A New Generation of Engine Combustion Processes for the Future?
© P. Duret (Editor) and Editions Technip, Paris, 2001, pp. 165 - 170
27 rue Ginoux, 75015 Paris

THE EUCAR VISION OF THE FUTURE OF POWERTRAINS

Rainer Aust
EUCAR SG Powertrain

ABSTRACT

Regarding the future of the automotive propulsion systems, we can state the following:

- The Internal Combustion Engine will be the dominating Propulsion System over the next 30 to 50 years
- Alternative Propulsion Systems will enrich the market, but they will not replace Diesel- or Gasoline-Engines
- There is a need for modified / synthetic fuels and lubricants, mainly sulphur, ash and aromatics free for Internal Combustion Engines (ICE)

This subordinates, that there is no revolutionary change in legislation and customer acceptance, the infrastructure for alternative fuels such as CNG, Methanol, Hydrogen, increase very slow because of costs.

The Internal Combustion Engines will develop themselves, which concern consumption and emissions. In future, there will be many different solutions and technologies with

- High sophisticated technologies for premium cars
- Compatible costs for small cars
- Combinations of different technologies

All with minimal emissions, strongly reduced fuel consumption, good driving performance and best efficiency from well to wheel.

Cars with very low fuel consumption are or will be available, it depends only on customers acceptance. Emissions of today's passenger cars are nearly zero, it is only about the first 30-40 seconds after start. The main fields for development are to reduce raw emissions, mainly in the first seconds, and to develop new complex aftertreatment systems with acceptable costs and lifespan.
To fulfil future emissions legislation, it is not done with developing engine technologies, for all these new technologies we need new modified/synthetic fuels and lubricants.

INTRODUCTION

Over hundred years, the internal combustion engines (Diesel and Otto) were the dominant drive for road vehicles. Engineers task was to increase power, speed and comfort.
Nowadays we are in a change-phase:

- environment aspects become more important and new technologies appear at the horizon.

In this presentation I would like to say something about:

- Achievements of internal combustion (IC) engines
- Technology trends for IC engines
- Fuels
- Future share of engines.

ACHIEVEMENTS

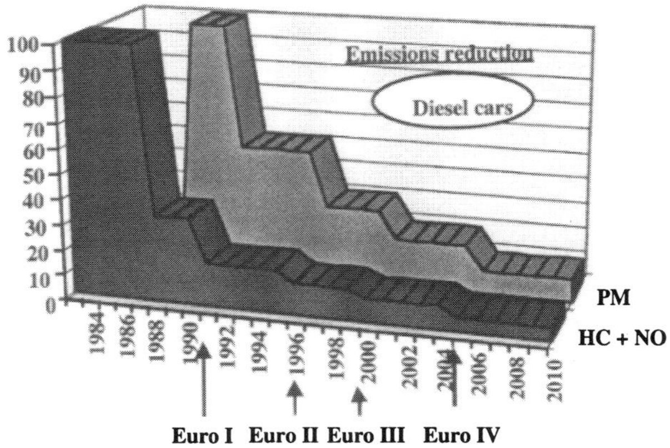

Source: European Commission, DG Enterprise, CAPoC5, Brussels 12 April

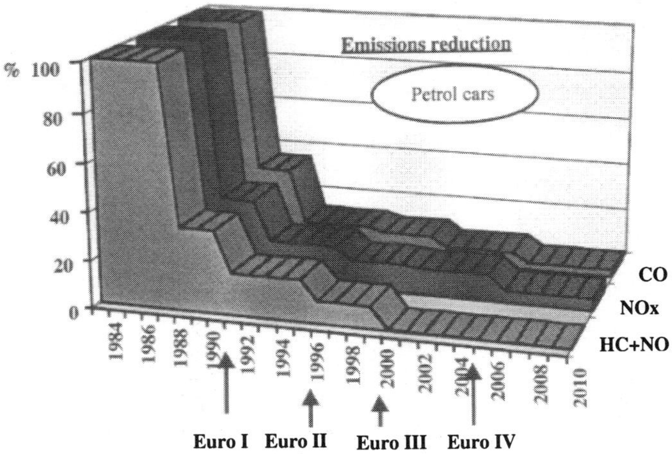

Source: European Commission, DG Enterprise, CAPoC5, Brussels 12 April

The statement is: regarding the Fleet Consumption Target and the achievements on emissions, in 14 years (1991 – 2005) from 100% to under 10%, there is no need for further emissions legislation, it increases only fuel consumption.

TECHNOLOGY TRENDS FOR IC ENGINES

The main tasks for Internal Combustion Engines are:

- Increase efficiency to reduce fuel consumption. One main technology is, to increase the in-cylinder-pressure through high-pressure charging.

- Development of complex aftertreatment systems. We have to consider that the regeneration of traps and catalysts increases fuel consumption.

- Reduction of raw emissions. Different technologies are in research and development.

Exhaust gas system diesel engine (principal example)

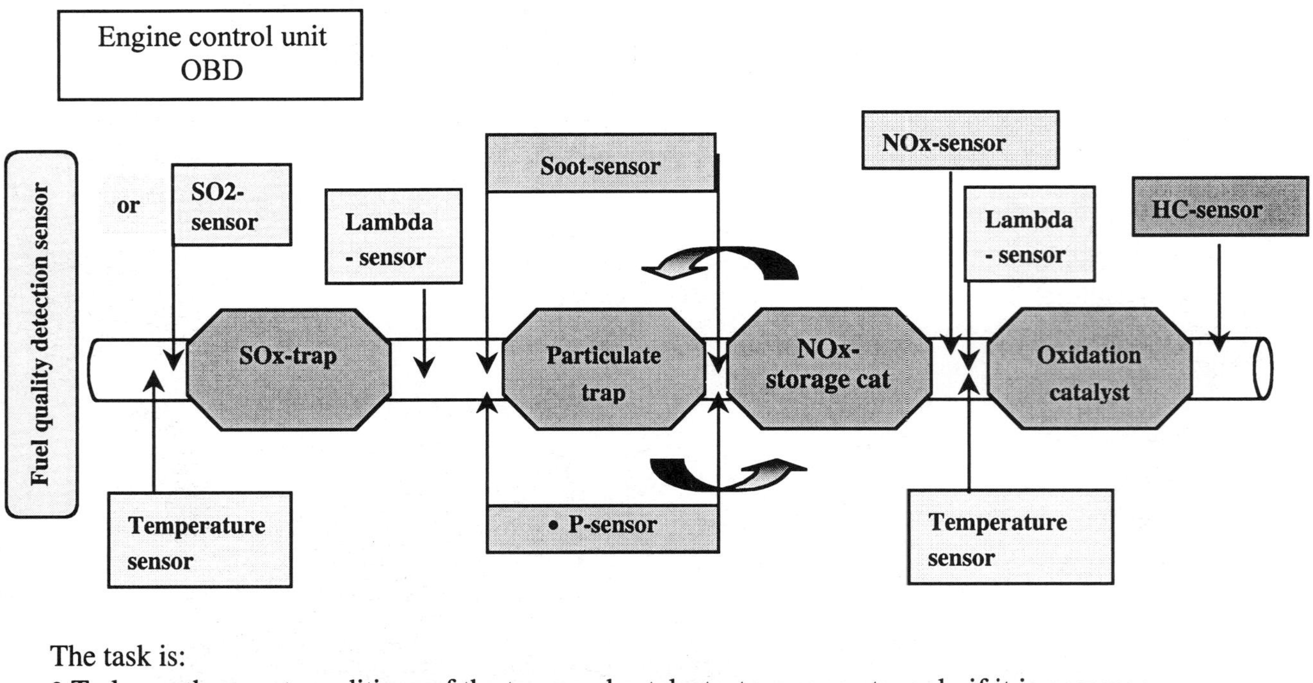

The task is:

- To know the exact conditions of the traps and catalysts, to regenerate only if it is necessary. Regeneration means to switch to rich operating conditions and that increases fuel consumption.
- To regard the whole chain from fuel/oil quality measurement, combustion process and raw emissions to the end of tailpipe.
- To define which sensors are really necessary or what could be managed by "virtual" sensors and software.

Multiple Injection Concept

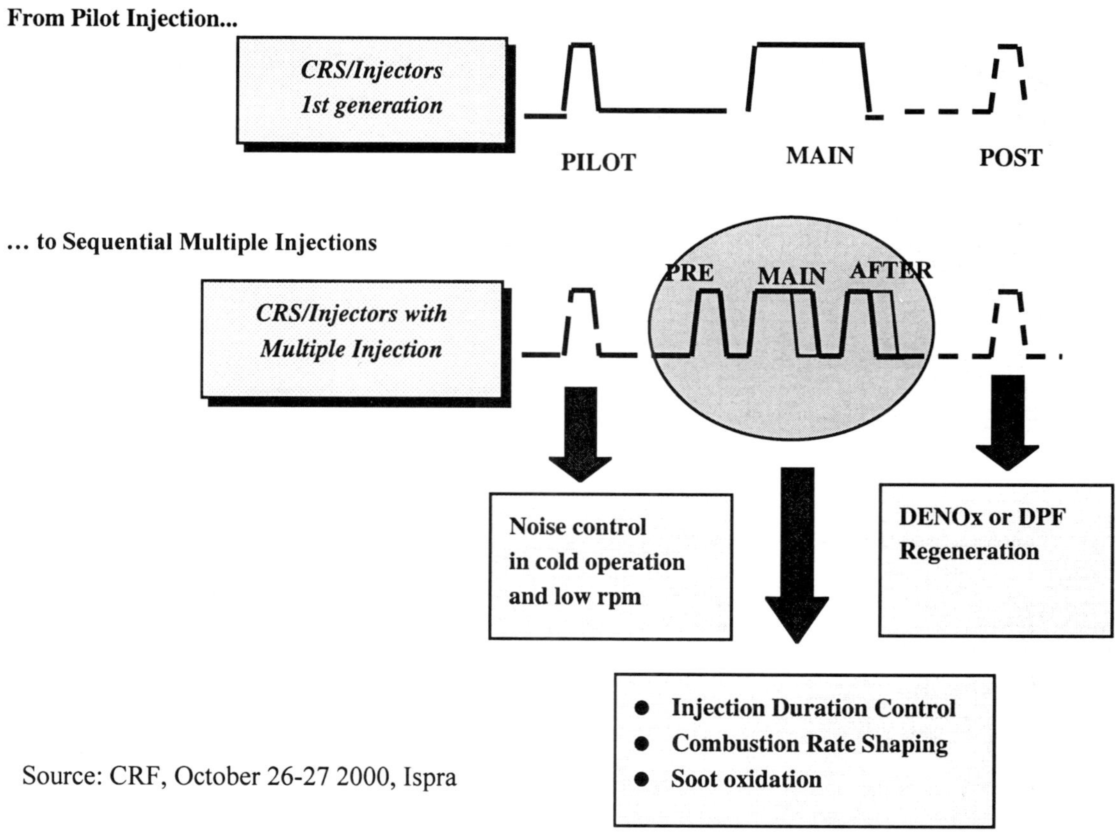

Source: CRF, October 26-27 2000, Ispra

Homogeneous Charge Compression Ignition (HCCI)
Controlled Auto Ignition (CAI)

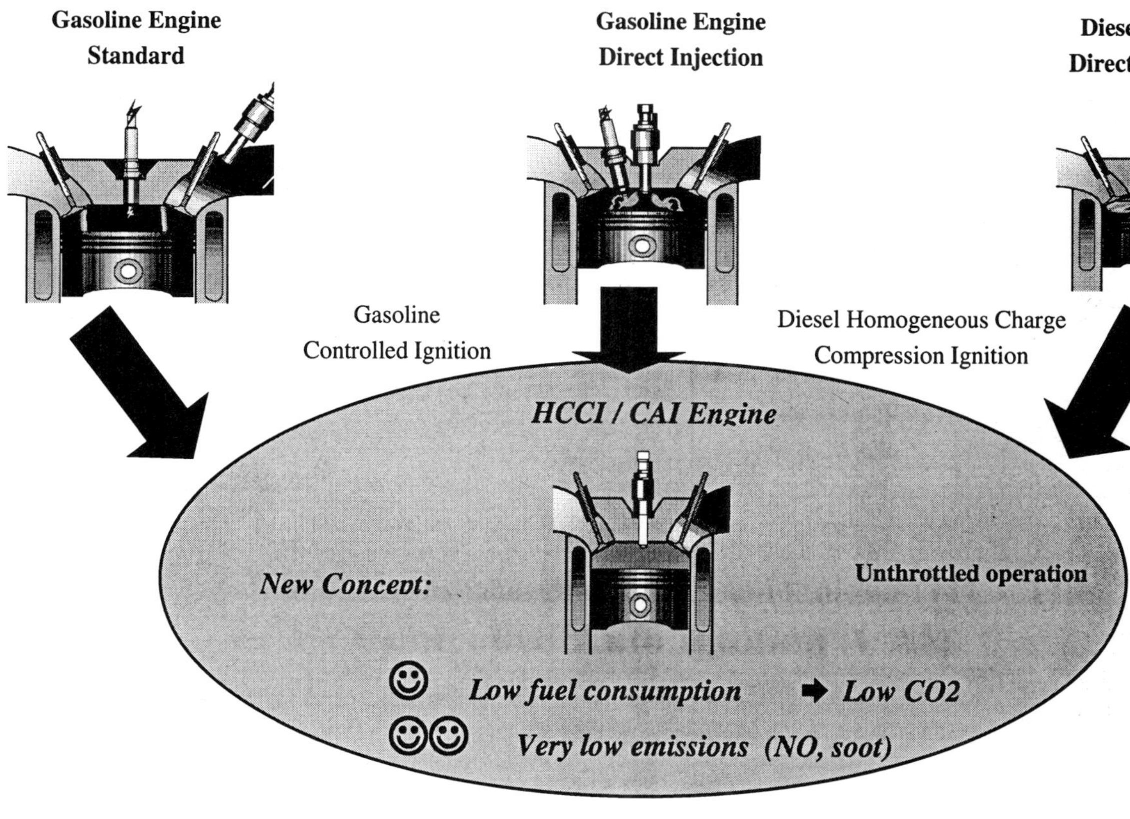

OUTLOOK ON FUELS

☞ In the next 20 to 30 years, the production of technically easily to extract conventional oil, will reach a maximum.

☞ The increasing need will satisfy then only through the inclusion of natural gas and non conventional deposits (heavy oil, tar sand, oil slate).

☞ The time of cheap oil is finally past (long-term price per barrel > 20$).

☞ The most economic alternative is medium term the production of synthetic fuels, first from natural gas, later from hard coal.
GTL-Diesel, GTL-Gasoline, Methanol and fuel components will increase.

☞ A hydrogen economy is very expensive, it is likely in the post-fossil time, if

☞ sufficiently renewable primary energies are available.

FUTURE SHARE OF ENGINES

Up to the year 2010 the picture is very clear:
- new technologies are in a research or pre-development status
- legislation is clearly defined

For the Internal Combustion Engine (ICE) there will be many different solutions and technologies with:
- high sophisticated technologies for premium cars
- compatible costs for small cars
- combinations of different technologies

All with:
- minimal emissions
- strongly reduced fuel consumption
- good driving performance

Alternative drive-train-systems have an insignificant market-share.

Beyond the year 2010, the share of engines depends mainly on:
- legislation
- availability of mature new technologies
- infrastructure of alternative fuels (CNG, Methanol, Hydrogen)
- availability of modified / synthetic fuels and lubricants
- costs and customer acceptance

For the Internal Combustion Engine (ICE) the diversity of solutions and technologies will continue.

Alternative drive-train-systems will increase their market-share

Subordinated that there is no revolutionary change in legislation and customer behavior, one can determine that the Internal Combustion Engines will have the dominant share of engines for the next decades.

Future Share of Engines in Europe (Passenger Cars)

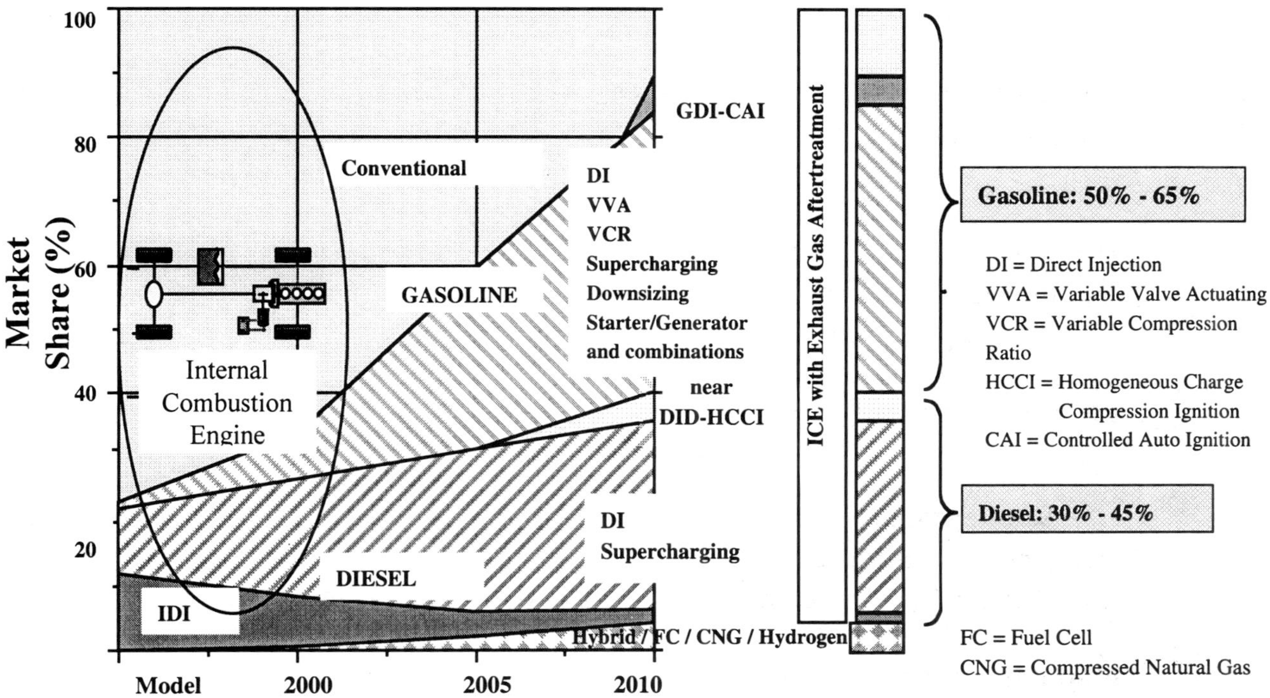

Source: VKA / RWTH Aachen, Graz, 12th International Congress Engine & Environment; EUCAR Conference, November 2000, Brussels

CONCLUSION

- The Internal Combustion Engine will be the dominating Propulsion System over the next 30 to 50 years, with best emissions and fuel consumption figures.
- There is a need for modified / synthetic fuels and lubricants ➢ mainly sulphur, ash and aromatics free
- Alternative Propulsion Systems will enrich the market, but they will not replace Diesel- or Gasoline-Engines

A New Generation of Engine Combustion Processes for the Future?
© P. Duret (Editor) and Editions Technip, Paris, 2001, pp. 171 - 174
27 rue Ginoux, 75015 Paris

CONTRIBUTION OF NEW GENERATION VARIABLE VALVE TRAINS AND ENGINE MANAGEMENT SYSTEMS FOR THE CONTROL OF NEW COMBUSTION PROCESSES

Jean Doreau
SAGEM

CONTROL OF THE COMBUSTION PROCESS

As for any type of process control the main functions are:

- observation of the key parameters (state variables)
- control through appropriate means (actuators)
- associated control strategies.

Most of the time those control strategies rely on physical models based on physical laws. Use of "fuzzy logics" is also considered.
Identification of models is the core of present investigations and as such have been extensively presented during this International Congress. This will enable an appropriate understanding of the leading parameters for efficiency and quality of combustion such as:

- air/fuel ratios
- temperature, pressure
- internal kinetics of gases and fuel (motion, tumble, swirl, ...)
- re-circulation rates of exhaust gases

Controls should in particular consider:

- limit condition such as "wall wetting" effects
- dynamic behaviour, instrumental for initialization phases

OBSERVATION MEANS

In depth access to the combustion will probably lead to complementing present external means (air flow mass, lambda sensors,...) by new technologies assessing the in-cylinder combustion behavior :

- ionization current
- in cylinder pressure sensors

ACTUATORS: FUELLING AND AIR MANAGEMENT

Engine management systems have been up to now driven by the "fuelling" side; an increasing emphasis on the "management of air" can now be expected.
For fuel introduction, quality of atomization will remain the instrumental factor, to be associated with higher dynamics.
Management of air and of the combustion gases will be needed to manage the conditions of appropriate initiation of combustion :

- flow
- temperature and pressure
- motion of gases in the combustion chamber

It is expected that fully variable control will become mandatory.

Following a general trend towards downsizing, the contribution of better controlled turbo-chargers, possibly "Electrically Assisted" will have to be considered.

FULLY VARIABLE VALVE ACTUATION

Full VVA will be necessary to control adequately the motion of air charge together with recycled burned gases and generate the mandatory conditions for homogeneous auto-ignition.

Of specific importance will be the operations in "multi-stroke modes", together with the potential of close loop combustion control.

Two main candidate technologies are considered:

- Electro-magnetic valve (EMV) as shown by the Figure 1 example,

Figure 1: SAGEM dual Electro-magnetic actuator

- Electro-hydraulic Valve Systems (EHVS) as shown by the Figure 2 example

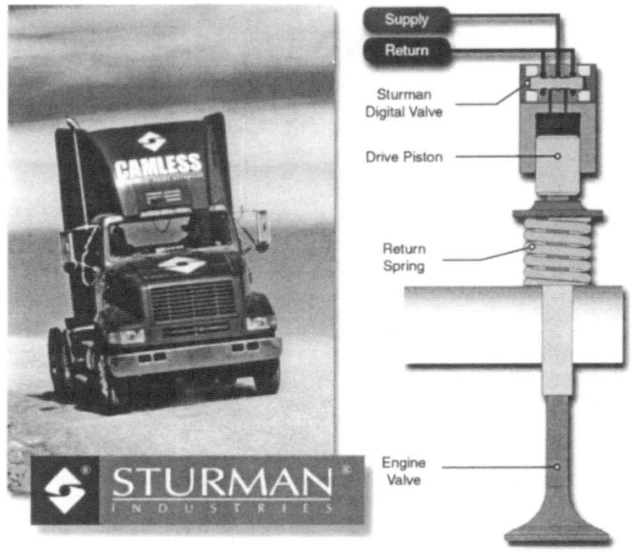

Figure 2: Sturman DHOS valve actuation module

The combination of EHVS with Electrically controlled , Hydraulically amplified Unit Injectors (HEUI) could be of specific interest for diesel engines.

COMPLEXITY OF CONTROLS

As indicated by the gasoline example of the following chart (Figure 3), New Combustion Processes (HCCI, CAI, ...) will introduce a major step in the complexity of the controls, as described by the number of parameters and variables to be managed.
Not only this will have a direct impact on the complexity of electronic architecture and application software, but also on the development methodologies of those very complex systems.

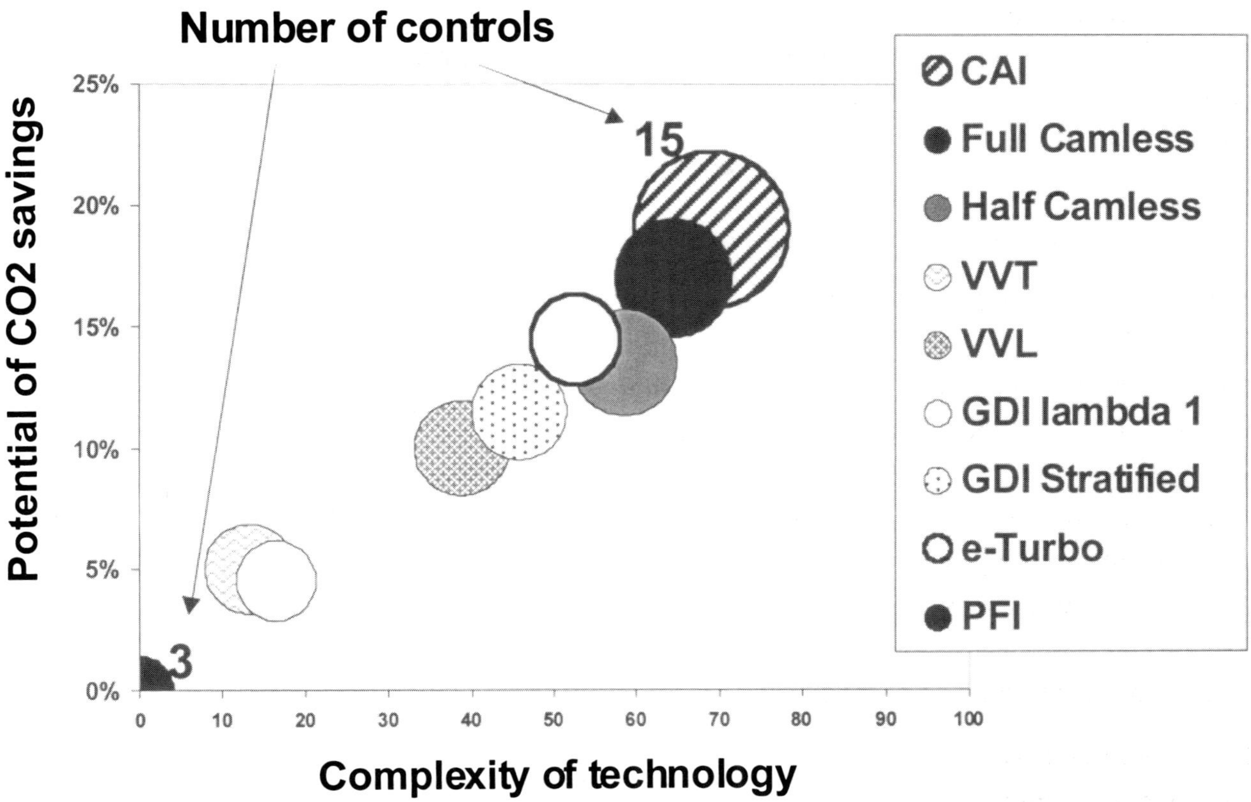

Figure 3: Complexity of controls and technologies (the gasoline engine example)

CONCLUSION:
DIESEL AND GASOLINE ENGINES ON PARALLEL ROAD MAPS

A trend is already acknowledged that Gasoline and Diesel EMS are converging towards common electronic architectures.

Management of Air will become an instrumental part of the new systems, also contributing to the synergies between Gasoline and Diesel.

It can also be anticipated that performances and fuel economy of those two types of engines, with the same emissions levels, will get closer.

A New Generation of Engine Combustion Processes for the Future?
© P. Duret (Editor) and Editions Technip, Paris, 2001, pp. 175 - 180
27 rue Ginoux, 75015 Paris

WHAT WILL BE THE FUTURE COMBUSTION AND FUEL-RELATED TECHNOLOGY CHALLENGES?

Xavier Montagne, Pierre Duret
IFP, France

ABSTRACT

Engine and fuels for transport are facing a double challenge:

- bring local pollution to the levels requested by W.H.O. city air quality standard,
- reduce CO_2 emission in order to minimize global warming risk.

These goals stimulate the development of both conventional and alternative engines and fuel technologies.

New combustion processes known as Controlled Auto-Ignition (CAI™) for gasoline engines and Homogeneous Charge Compression Ignition (HCCI) for diesel engines are the subjects of intensive research world-wide and particularly at IFP. Because of the thermo-chemical conditions of the charge, the thermal NOx formation is, in principle, much lower than with the typical flames of conventional engines. Indeed, these new combustion processes bring down NOx to very low level without post-treatment, while maintaining a very high thermodynamic efficiency of the combustion.

The control of such combustion is complex due to the number of technological and fuels parameters to take into account. Considering fuels, it appears that their formulation could have an important impact on the optimal operating ranges of both CAI or HCCI. Then, tailored fuel formulation, based on a close selection of qualities of combustion initiation, vapor formation, blend preparation, interaction with residual gas and flame diffusion, will strongly contribute to the success of these new combustion processes. The targets of these new combustion modes induce to consider new key parameters for fuels. For example, fuel parameters, such as the octane of the gasoline or the cetane of the diesel fuel, which have key values for conventional combustion might not have similar value for CAI and HCCI combustion. One further view could even raise the question of one unique fuel for a unique combustion process.

INTRODUCTION

Car manufacturer's and refiner's new challenges are to evolve technologies to improve significantly the air quality through a strong decrease in the exhaust gas release and to reduce CO_2 emission to control the greenhouse effect. Thus, direct injection engines, either diesel or spark ignited, underwent strong developments during the past 5 years and gained efficiency but there is still a need for a large reduction in their NOx and/or particulate emissions to pass the future Euro 4, Euro 5, and further regulations. Nevertheless, a strategy of more and more complex exhaust gas aftertreatment is possible but, at present, it is still associated with quite important concerns such as its cost, its sensibility to sulphur, ... which rises some unanswered questions for after 2005. It is the reason why important new developments, both for fuels and engines, are necessary as illustrated in figure 1 [1].

New combustion processes are under investigation for their potential to reduce to very low levels the NOx production. They constitute alternative or complementary answers to the sophisticated and complex after-treatment strategy. Controlled Auto-Ignition (CAI), Homogeneous Charge Compression Ignition (HCCI) and a number of acronyms designate these combustion processes. Their principle consists in preparing a highly diluted mixture, and to obtain its simultaneous ignition in the whole space of the combustion chamber while precisely controlling this combustion for best performance in terms of efficiency and NOx emissions (timing and heat release rate during the cycle). Because of the thermo-chemical characteristics of the charge, the thermal NOx formation and the soot production are in principle much lower than with either diffusion flame, typical of the conventional diesel, or with the propagation flame, typical of the conventional spark ignition engine.

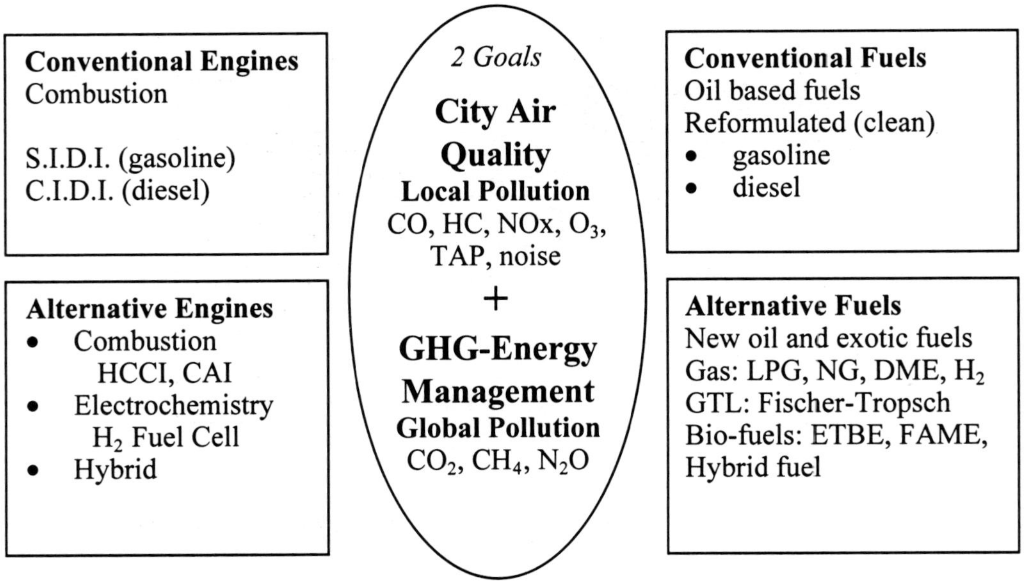

Figure 1: Overview of engine and fuel technologies for future environmental goals.

Then, the practice and the development of these new combustion processes are major axis for research and progress. Presently available results indicate that large technological improvements can be expected. Further, a strong impact of the fuel quality can be foreseen and tailored formulations could be required.

This paper purpose is to point out the formulation requirements of fuels when specially designed to improve the operating range of engines with these new combustion processes. The final goal being that, both gasoline CAI and diesel HCCI, combined with an optimized fuel, fully reveal their strong potential for practical use in the production power trains of the passenger cars for the end of this decade.

BACKGROUND ON FUEL AND ADDITIVE EFFECT ON CAI COMBUSTION

An international consortium project, involving several car manufacturers and fuel companies was undertaken by IFP during 1995-1997 [2]. The project focused on the effects of fuels and additives on CAI combustion. Its main purposes were as follows:

- ❑ to experimentally define and validate (using a 2-stroke engine as experimental tool) a specific engine test procedure, giving relevant CAI criteria towards fuel characteristics;
- ❑ to analyze and compare different fuels (with various formulations), according to the procedure developed;
- ❑ and to evaluate the possibility to enlarge the CAI range with "CAI-designed" fuels and without any detrimental effect on knock.

Figures 2 and 3 show some examples of the results obtained by this project.

Figure 2 compares the Controlled Auto-Ignition Number (also named CAN index) measured for various fuel products. The CAN index was defined taking into account the speed and load range when the engine run in CAI with the spark ignition turned off. The wider the CAI speed and load range, the higher is the CAN index.

In Figure 2 results, the CAN index varies from about –20 to about +45. Even if details on fuel formulations are not given in the figure, it is already possible to notice that fuel formulations including oxygenated products generally have higher CAN index than non-oxygenated fuels. What it is very spectacular, is the best situation results from a synergy between 2 oxygenated components with very contrasting ignition characteristics.

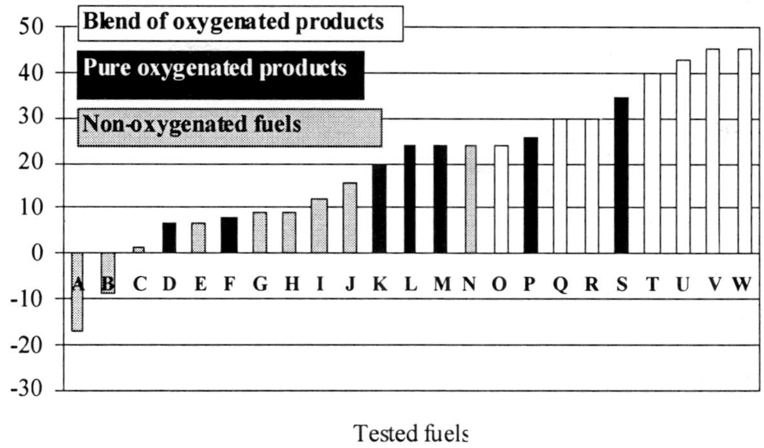

Figure 2: Controlled Auto-Ignition Number (CAN) for various fuel products.

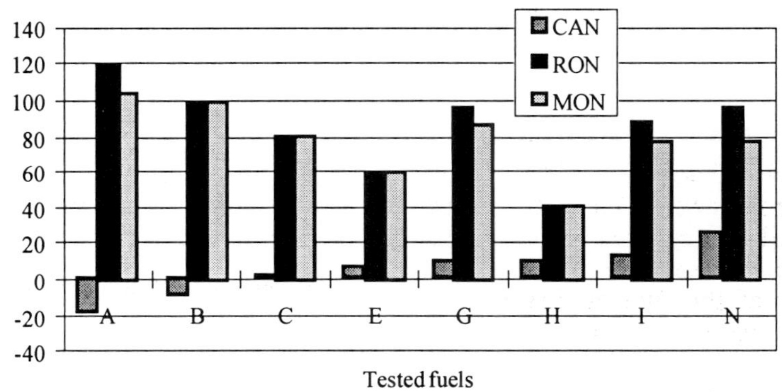

Figure 3: Comparison of CAN, RON and MON for various fuels.

Figure 3 shows that for eight fuels with different Research Octane Number (RON) and Motor Octane Number (MON), the CAN index varies without any obvious correlation with the RON and MON values. It can even be noted that the highest CAN index, which indicates the fuel giving the widest range of Controlled Auto-Ignition, is obtained for one of the fuels having the highest RON and MON which means the strongest knock resistance. Then, we notice that RON and MON parameters, generally recognized as very relevant to evaluate the potential resistance to knock of a gasoline fuel, are no more relevant to characterize its adaptation to CAI combustion.

Considering these two figures as example, one can understand that the fuel formulation could have a significant impact on the performance of future CAI engines and that we are just at the beginning of the investigations on this fuel effect on the control of the CAI combustion.

NEW AXIS FOR FUEL FORMULATION

We have demonstrated how a tailored fuel formulation approach could optimize the benefits of the CAI combustion process. Yet, at first, CAI and HCCI processes will have to be studied separately.

The CAI process lies on a controlled auto-ignition of the load and is associated with a more or less wide operating range. The target will be to enlarge as much as possible this range to optimize the benefits of the combustion process. This widening of the favorable running conditions depends strongly on the type of products that can be found in the combustion chamber, either directly linked with fuel formulation or formed by combination with residual gas. Then, as it has been demonstrated in studies on the ATAC combustion in 2 stroke engines [3], the measurement of the hydrocarbons found in the exhaust gas enable to state the behavior of the fuel basis during the combustion process. To reach the optimal operation of these new processes, it is necessary to take into account a large number of parameters such as the ignition process (by spark, temperature, ignition delay, active species,...), the internal fluid dynamic control (which controls the stratification between the fuel, the fresh charge, the internal and external EGR), the combustion development process (by flame propagation, flame diffusion, by simultaneous auto-ignition and combustion development in the whole space of the combustion chamber,...). When running gasoline CAI modes, each of these parameters depends strongly on the fuel characteristics.

During all the combustion initiation phase the chemical type of species and their volatility are major criteria. Besides, octane numbers, which are the parameters usually considered to control this phase in spark ignition engines, appear to have in these new conditions a very peculiar interest and some species with low octane numbers stand as very promising products [2].

The second step of combustion depends on chemical properties. During this phase, the blending of fresh gas and exhaust gas has to generate reactive species, enabling the combustion to progress without any help from an external ignition. As this phase seems to be the more important and as shown in [2], the choice of fuel bases will be very helpful. This process correct development can be favored by the selection of refinery basis or of products issued from petrochemistry, especially oxygenated compounds. Today different axis to select fuels with interesting reactivity and ignition qualities are identified, showing de facto that level of octane numbers using for EN228 gasoline are not typically the best here.

Lastly, the fuel physical characteristics are key parameters for flame propagation on the combustion itself. Fuels have also to be prepared considering these needs.

The problematic is not the same for the HCCI mode but the approach is very similar to the one used to improve the operation of diesel engines in HCCI mode. The homogeneous combustion of diesel fuel immediately faces 2 major issues which are the fuel vaporization and the combustion initiation. Thus, the fuel formulation has to be oriented around these 2 axis to closely control the development of the blend formation and the combustion initiation, including interactions with exhaust gas. When running in HCCI mode, having a fast combustion initiation does not appear as appropriate. Then, one can expect that a high cetane number will no more be a key parameter for fuel formulation contrary to have a fast vaporisation.

Therefore, the optimization of the combination of engine combustion technologies and fuel physico-chemical parameters is the way to obtain engines running in CAI or HCCI modes with a correct operating range as wide as possible.

CONCLUSIONS

The new challenges for the automotive industry and refiners concern the availability of new engines and fuels to reach the more and more severe regulations and to reduce CO2 emissions. In this context, the new combustion modes of engines, as CAI and HCCI modes represent important progress axis. As previously shown, technological development are determining to reach the more ambitious objectives. Nevertheless, the great potential of these new combustion modes should be further enhanced with optimised fuels. Preliminary tests have shown that new formulation of fuels give very promising results and it appears that some characteristics of fuels emerge as key parameters. In an other hand, for example, it is clear that the optimal values of parameters as octane number of gasoline or cetane number of diesel fuel will not have same impact for CAI and HCCI combustion.

Finally, there will be in the future two main ways of engine & fuels research & development on the subject of new combustion processes. One can start from the auto-ignition of gasoline and the other one from the partial homogeneous ignition of diesel. The potential of both will have to be explored with more and more appropriate fuel specifications. At the end, these two ways could perhaps converge towards a single fuel for a single combustion process (such as the idea of the "Combined Combustion System" described in [4]). This could be a major challenge of a future generation of engines and energy technologies.

REFERENCES

[1] P. Duret, A. Douaud, New Combustion Processes: Homogeneous Diesel and Auto-Ignition Gasoline Towards Zero Emissions

with Major Engine and Fuel Technology Challenges, Global Powertrain Congress, June 5-7, 2001

[2] J. Lavy, P. Duret - IFP International Consortium Project on the Effects of Fuels and Additives on Controlled Auto-Ignition - Consortium reports 1995-97

[3] A. Petit, J. Lavy, G. Monnier and X. Montagne - Speciated Hydrocarbon Analysis: A Helpful Tool for Two-Stroke Engine Development - IFP International Seminar 1993.

[4] W. Steiger - Volkswagen Strategy to High Efficient Powertrains - 22th International Vienna Motor Symposium - 26 & 27 April 2001

Achevé d'imprimer
Impression : EUROPE MEDIA DUPLICATION S.A.
F 53110 Lassay-les-Châteaux
N° 8914 - Dépôt légal : novembre 2001
N° d'éditeur 1079